Advanced Courses in Mathematics
CRM Barcelona

Centre de Recerca Matemàtica

Managing Editor:
Manuel Castellet

Noel Brady
Tim Riley
Hamish Short

The Geometry of the
Word Problem for Finitely
Generated Groups

Birkhäuser Verlag
Basel · Boston · Berlin

Authors:

Noel Brady
Department of Mathematics
Physical Sciences Center
601 Elm Ave
University of Oklahoma
Norman, OK 73019
USA
e-mail: nbrady@math.ou.edu

Tim Riley
Department of Mathematics
310 Malott Hall
Cornell University
Ithaca, NY 14853-4201
USA
e-mail: tim.riley@math.cornell.edu

Hamish Short
Centre de Mathématiques et Informatique
Université de Provence
39 rue Joliot Curie
13453 Marseille cedex
France
e-mail: hamish.short@cmi.univ-mrs.fr

2000 Mathematical Subject Classification: 20F65, 20F67, 20F69, 20J05, 57M07

Library of Congress Control Number: 2006936541

Bibliografische Information Der Deutschen Bibliothek
Die Deutsche Bibliothek verzeichnet diese Publikation in der Deutschen Nationalbibliografie; detaillierte
bibliografische Daten sind im Internet über <http://dnb.ddb.de> abrufbar.

ISBN 978-3-7643-7949-0 Birkhäuser Verlag, Basel – Boston – Berlin

© 2006 Birkhäuser Verlag, P.O. Box 133, CH-4010 Basel, Switzerland
Part of Springer Science+Business Media
Printed on acid-free paper produced from chlorine-free pulp. TCF ∞
Printed in Germany
ISBN-10: 3-7643-7949-9
ISBN-13: 978-3-7643-7949-0
9 8 7 6 5 4 3 2 1

∞   e-ISBN-10: 3-7643-7950-2
e-ISBN-13: 978-3-7643-7950-6

www.birkhauser.ch

Contents

II   Filling Functions
Tim Riley   **81**

1   Filling Functions   **89**

2   Relationships Between Filling Functions   **109**

3   Example: Nilpotent Groups   **123**

4   Asymptotic Cones   **129**

Foreword

The advanced course on *The geometry of the word problem for finitely presented groups* was held July 5–15, 2005, at the Centre de Recerca Matemàtica in Bellaterra (Barcelona). It was aimed at young researchers and recent graduates interested in geometric approaches to group theory, in particular, to the word problem. Three eight-hour lecture series were delivered and are the origin of these notes. There were also problem sessions and eight contributed talks.

The course was the closing activity of a research program on *The geometry of the word problem*, held during the academic year 2004–05 and coordinated by José Burillo and Enric Ventura from the Universitat Politècnica de Catalunya, and Noel Brady, from Oklahoma University. Thirty-five scientists participated in these events, in visits to the CRM of between one week and the whole year. Two weekly seminars and countless informal meetings contributed to a dynamic atmosphere of research.

The authors of these notes would like to express their gratitude to the marvelous staff at the CRM, director Manuel Castellet and all the secretaries, for providing great facilities and a very pleasant working environment. Also, the authors thank José Burillo and Enric Ventura for organising the research year, for ensuring its smooth running, and for the invitations to give lecture series. Finally, thanks are due to all those who attended the courses for their interest, their questions, and their enthusiasm.

Part I

Dehn Functions and Non-Positive Curvature

Noel Brady

Preface

In this portion of the course we shall explore some ways of constructing groups with specific Dehn functions, and we shall look at connections between Dehn functions and non-positive curvature. The presentation of the material will proceed via a series of concrete examples. Further, each section contains exercises.

Relevant background topics from the topology of groups (such as *graphs of groups* and *graphs of spaces*), and from non-positively curved geometry (such as *CAT(0) spaces* and *CAT(0) groups*, and *hyperbolic groups*) are introduced with a view to the immediate applications in this course. So we shall learn definitions and statements of the major results in these areas, and proceed to examples and applications rather than spending time on proofs. Here is an outline of how we shall proceed.

First, we study the *snowflake construction*, which produces groups with Dehn functions of the form x^α for a dense set of exponents $\alpha \geqslant 2$, including all rationals. These groups and constructions are far from the non-positively curved universe; for instance, the snowflakes are not even subgroups of the non-positively curved groups mentioned in the next paragraph.

The next series of examples are all subgroups of non-positively curved groups. The non-positively curved groups in question are *CAT(0) groups* and *hyperbolic groups*. Subgroups of non-positively curved groups are not well understood at present; the collection of subgroups is potentially a vast reservoir of new geometries and groups. One key difficulty in this field is that there is a real dearth of concrete examples. Another problem is that there are very few good tools for analyzing the geometry of subgroups of non-positively curved groups.

We begin by examining a construction for embedding certain amalgamated doubles of groups into non-positively curved groups that has its foundations in a paper of Bieri. As an application, we construct a family of CAT(0) 3-dimensional cubical groups which contain subgroups with Dehn functions of the form x^n for each $n \geqslant 3$. The groups that are being doubled are free-by-cyclic groups which are the fundamental groups of non-positively curved squared complexes. We define *Morse functions* on affine cell complexes, and use Morse theoretic techniques to see that the fundamental groups of the squared complexes above are indeed free-by-cyclic.

The Morse theory techniques are applied to non-positively curved cubical complexes for the remaining applications and examples. In one application we look at Morse functions on cubical complexes corresponding to *right-angled Artin groups*. The Artin group is the fundamental group of the associated cubical complex, and the circle-valued Morse function induces an epimorphism from the Artin group to $\mathbb{Z}$. The geometry of the kernel of this epimorphism is intimately related to the geometry and topology of the *level sets* of a lift of the Morse function to the universal cover. As examples, we produce right-angled Artin groups containing subgroups which have Dehn function of the form x^n for $n \geqslant 3$. These examples have a very different feel to the embedded doubled examples above. In the doubled examples, the Dehn function exponent is closely related to the *distortion* of free subgroups in the doubled group. This is not the case with the right-angled Artin examples.

As a final example, we construct a branched cover of a 3-dimensional cubical complex, with the following properties. The fundamental group is hyperbolic. There is an epimorphism to $\mathbb{Z}$ whose kernel is finitely presented but not hyperbolic. The kernel is known not to be hyperbolic because it is not of type F_3; an explicit calculation of its Dehn function is yet to be carried out.

Morse theory is the major background theme in this portion of the course. It is used explicitly in the later sections on Artin groups and on branched covers. It is used to recognize free-by-cyclic groups in the section on embedding doubles. It is also the motivation for the *torus construction* which produces the vertex groups in the graph of groups description of the snowflake groups. The torus construction leads to a whole range of groups with interesting geometry and topology. These include a famous example due to Stallings of a finitely presented group which is not of type F_3. The torus construction leads to quick descriptions for a range of variations of Stallings' example, some of which have cubic Dehn functions. Some may have a quadratic Dehn function. There is much to explore here.

Many people have contributed in different ways to the preparation of these lectures. I acknowledge the contributions of coauthors whose joint projects form the basis for various sections of these lectures; Josh Barnard, Mladen Bestvina, Martin Bridson, Max Forester and Krishnan Shankar. I thank Jose (Pep) Burillo and Enrique Ventura for organizing the concentration year on the Geometry of the Word Problem, and for inviting me to participate. Thanks are also due to Hamish Short and Tim Riley, who also spoke at the mini-course on the Geometry of the Word Problem, and who offered comments on the lectures. I thank Laura Ciobanu and Armando Martino for helpful comments and words of encouragement during the early stages of writing these notes. Finally, many thanks are due to all at the Centre de Recerca Matemàtica in Barcelona for their excellent professional support and for providing a very pleasant working environment.

Chapter 1

The Isoperimetric Spectrum

In this chapter we focus on one aspect of the theory of Dehn functions; namely the question which functions of the form x^α are Dehn functions of finitely presented groups. We can ask about the range of exponents $\alpha \in [1, \infty)$ such that x^α is the Dehn function of a finitely presented group. Since there are only countably many isomorphism classes of finitely presented groups, this is a countable collection of real numbers in $[1, \infty)$. We call this collection of real numbers the *isoperimetric spectrum*.

Sections in this chapter are organized as follows. The definition of the IP spectrum and a survey of results, the definition of Perron–Frobenius eigenvalues and the statement of the main theorem are provided in the first section. The second section covers relevant topological background; graphs of spaces and graphs of groups, the torus construction and the definition of vertex groups. In Section 1.3 we give two illustrative examples of snowflake groups, then define the general snowflake groups and sketch lower bounds arguments for their Dehn functions. The next subsection gives the sketch of the upper bound arguments. In the fourth section we discuss open questions and possible research directions.

1.1 First order Dehn functions and the isoperimetric spectrum

1.1.1 Definitions and history

In this section we define the isoperimetric spectrum, $\mathbb{P}$, and give some history of the results concerning the structure of $\mathbb{P}$. The main point is that the gap between 1 and 2 in $\mathbb{P}$ corresponds to the deep and useful characterization (due to Gromov) of hyperbolic groups as those with sub-quadratic isoperimetric functions.

Definition 1.1.1 ($\mathbb{P}$-Spectrum). A real number α is said to be an *isoperimetric exponent* if there exists a finite presentation with Dehn function $\delta(x) \sim x^{\alpha}$. The collection of all isoperimetric exponents is called the *isoperimetric spectrum* and is denoted by $\mathbb{P}$.

Remark 1.1.2. By definition of equivalence of functions, we can assume that isoperimetric exponents lie in the set $[1, \infty)$. Since there are countably many finite presentations, $\mathbb{P}$ is a countable subset of $[1, \infty)$.

A basic question concerning isoperimetric inequalities of groups is to determine the structure of $\mathbb{P}$. The main reason people are interested in this is because of the following remarkable theorem of Gromov.

Theorem 1.1.3 (Sub-quadratic is hyperbolic). *The following statements are equivalent for a finitely presented group G.*

1. *G has a sub-quadratic isoperimetric inequality.*

2. *G has a linear isoperimetric inequality.*

3. *G is a hyperbolic group.*

This theorem implies that there is a gap in $\mathbb{P}$ between 1 and 2. The gap corresponds to the sub-quadratic reformulation of hyperbolicity for groups. This sub-quadratic criterion has been used to prove useful theorems about hyperbolic groups, such as the Bestvina–Feighn Combination Theorem.

So people were led to ask if there are other gaps in $\mathbb{P}$, and if so, whether these gaps had any algebraic or geometric significance for groups. Figure 1.1 gives an overview of the history of discoveries about $\mathbb{P}$.

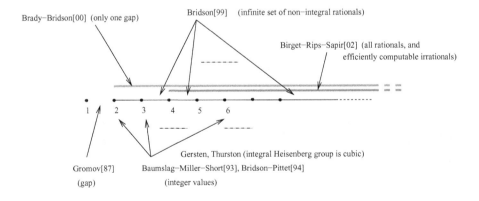

Figure 1.1: History of discoveries about the isoperimetric spectrum.

Gromov [29] described the intuition behind the sub-quadratic characterization of hyperbolicity in his seminal paper "Hyperbolic Groups". Detailed proofs of this characterization were given by Bowditch [7], Ol'shanskii [32], and Papasoglu [33]. S. M. Gersten [27] and W. Thurston [25] gave arguments to show that the integral Heisenberg group has a cubic Dehn function. Then Baumslag–Miller–Short [3], and later Bridson–Pittet [18] found groups with arbitrary integral isoperimetric exponent. Bridson [13] combined nilpotent groups in various ways to give an infinite family of groups with non-integral, rational isoperimetric exponents. This family of fractions is far from dense; there were still many gaps in $\mathbb{P}$ at this stage.

The next two results demonstrated that there are no gaps in the $[2, \infty)$ portion of $\mathbb{P}$. One result, due to Brady–Bridson [9], consists of a family of finitely presented groups whose isoperimetric exponents include a dense collection of transcendental numbers in $[2, \infty)$. This proved that there is only one gap in $\mathbb{P}$. The other results, due to Sapir–Birget–Rips [34] and Birget–Ol'shanskii–Rips–Sapir [6] gave much more detailed information about $\mathbb{P}$ in the range $[4, \infty)$. For example, if a real number $\alpha > 4$ is such that there is a constant $C > 0$ and a Turing machine which calculates the first m digits of the decimal expansion of α in time at most $C 2 2^{2^{Cm}}$, then $\alpha \in \mathbb{P}$. Furthermore, if $\alpha \in \mathbb{P}$, then there exists a Turing machine which computes the first m digits of the decimal expansion of α in time bounded above by $C 2 2^{2^{2^{Cm}}}$. Indeed they gave much more detailed information about the posssible types of Dehn functions (not necessarily power functions) which are bounded below by x^4.

1.1.2   Perron–Frobenius eigenvalues and snowflake groups

We do not have time to do justice to the deep and powerful techniques of Birget–Rips–Ol'shanskii–Sapir or the more recent work of Sapir–Ol'shanksii in this short course. Instead, we shall focus on giving a detailed description of the groups recently produced by Brady–Bridson–Forester–Shankar. We shall sketch the different techniques involved in proving lower and upper bounds for their Dehn functions.

The groups $G_{r,P}$ developed by Brady-Bridson-Forester-Shankar are best described as graphs of groups with right-angled Artin vertex groups and infinite cyclic edge groups. Their definition starts with an irreducible integer matrix P and a rational number r which is greater than all of the row sums of P. The underlying graph for the graph of groups description of $G_{r,P}$ has transition matrix equal to P. We begin by reviewing definitions and properties of irreducible matrices and transition matrices.

Definition 1.1.4 (Irreducible matrix). An $(R \times R)$-nonnegative matrix P is *irreducible* if for every $i, j \in \{1, \ldots, R\}$ there exists a positive integer m_{ij} such that the ij-entry of $P^{m_{ij}}$ is positive.

The main result about irreducible matrices is the following theorem of Perron–Frobenius.

Theorem 1.1.5 (Perron–Frobenius). *Suppose P is an irreducible, non-negative $(R \times R)$-matrix. Then there exists an eigenvalue λ such that:*

1. *λ is real and positive,*

2. *λ has a strictly positive eigenvector, and the λ-eigenspace is 1-dimensional,*

3. *if μ is another eigenvalue of P, then $|\mu| < \lambda$,*

4. *λ lies between the maximum and the minimum row sums of P, and λ is equal to this maximum or minimum only when all row sums are equal. Likewise for column sums.*

Definition 1.1.6 (Perron–Frobenius eigenvalue). The eigenvalue λ in the Perron–Frobenius theorem above is called the *Perron–Frobenius eigenvalue* of the matrix P.

We now recall the definitions of graph and of transition matrix associated to a graph.

Definition 1.1.7 (Graph). A *graph* Γ consists of a pair of sets $(E(\Gamma), V(\Gamma))$ and maps $\partial_\iota, \partial_\tau : E(\Gamma) \to V(\Gamma)$ and an involution $: E(\Gamma) \to E(\Gamma) : e \mapsto \bar{e}$ such that $e \neq \bar{e}$ and $\partial_\iota \bar{e} = \partial_\tau e$ for all $e \in E(\Gamma)$.

You can think of elements of $V(\Gamma)$ as *vertices*, and elements of $E(\Gamma)$ as *oriented edges*. The oriented edge e has *terminal vertex* $\partial_\tau(e)$ and *initial vertex* $\partial_\iota(e)$.

Definition 1.1.8 (Transition matrix of a directed graph). Let Γ be a finite, directed graph with vertex set $\{v_1, \ldots, v_R\}$. The *transition matrix* of Γ is an $(R \times R)$-matrix P such that P_{ij} equals the number of directed edges from vertex v_i to v_j.

Example. For the first example below, determine the transition matrix, and for the second example, determine a directed graph whose transition matrix is the given matrix.

1. Graph to matrix.

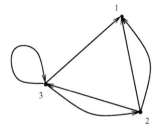

2. Matrix to graph.

$$P = \begin{pmatrix} 1 & 2 \\ 3 & 4 \end{pmatrix}$$

3. Suppose that the transition matrix of a finite directed graph is irreducible. What does this say about paths in the graph? (What do the entries of P^2 count in the graph?)

We are now ready to state the main result of this chapter.

Theorem 1.1.9 (Brady–Bridson–Forester–Shankar). *Let P be an non-negative, irreducible square matrix with integer entries, and having Perron–Frobenius eigenvalue $\lambda > 1$. Let r be a rational number which is greater than the largest row sum of P. There exists a finitely presented group $G_{r,P}$ with Dehn function $\delta(x) \sim x^{2\log_\lambda(r)}$.*

Remark 1.1.10 (Snowflake Groups). The groups $G_{r,P}$ of Theorem 1.1.9 above are called *snowflake groups*. They will be defined precisely in Section 1.3.1 below. This terminology will become apparent in the sketch of the lower bounds for their Dehn functions.

Remark 1.1.11. For any pair of positive integers $a < b$, we can take P to be the 1×1 matrix (2^a), and r to be 2^b. Then the snowflake group $G_{r,P}$ has Dehn function $\delta(x) \sim x^{2b/a}$. Thus, $\mathbb{P}$ contains all the rational numbers in $[2, \infty)$.

We postpone a formal definition of the snowflake groups for a few subsections. Instead, we give a first level description of the snowflake groups. The matrix P is the transition matrix of a finite directed graph Γ. The snowflake group $G_{r,P}$ is the fundamental group of a graph of groups, whose underlying graph is Γ, whose edge groups are all $\mathbb{Z}$. The vertices of Γ are in one-to-one correspondence with the rows of P, and the ith vertex group V_{m_i} is defined below, and depends on the integer m_i which is the ith row sum of the matrix P. The rational number r and the directed edges all encode how to map the infinite cyclic edge groups into these vertex groups.

We shall briefly review graphs of groups and graphs of spaces, then describe the vertex groups and list their properties, before giving a detailed description of the snowflake groups.

1.2 Topological background

In this section we describe two topological constructions which are key to the definition of the snowflake groups. The first is the notion of a graph of spaces and the corresponding notion of a graph of groups. The snowflake groups are defined to be very special graphs of groups. The second notion is the *torus construction*. This is used to define the vertex groups in the graph of groups definition of the snowflake groups. The torus construction is interesting in its own right, and has particular relevance to kernel subgroups of right-angled Artin groups. The torus construction will appear later in the examples in Section 2.5.

1.2.1   Graphs of spaces and graphs of groups

The snowflake groups are defined as graphs of groups, and the vertex groups in this description are in turn defined as graphs of $\mathbb{Z}^2$ groups with $\mathbb{Z}$ edge groups. We begin by defining graphs, and graphs of groups and graphs of spaces.

Definition 1.2.1 (Graph of spaces). A *graph of spaces* consists of a finite graph Γ, a vertex space X_v associated to each vertex $v \in V(\Gamma)$, an edge space X_e associated to each edge $e \in E(\Gamma)$, and continuous maps $f_{\iota,e} : X_e \to X_{\iota(e)}$ and $f_{\tau,e} : X_e \to X_{\tau(e)}$ for each edge e of Γ.

The *total space of the graph of spaces* above is defined as the quotient space of the disjoint union

$$\left(\bigcup_{v \in V(\Gamma)} X_v \right) \cup \left(\bigcup_{e \in E(\Gamma)} X_e \times [0,1] \right)$$

by the identifications $(x, 0) \sim f_{\iota(e)}(x)$ for all $x \in X_e$ and $(x, 1) \sim f_{\tau(e)}(x)$ for all $x \in X_e$.

The *fundamental group of the graph of spaces* above is defined to be the fundamental group of the total space.

Remark 1.2.2. Given a group G one can consider the presentation 2-complex K_G corresponding to a presentation of G.

Definition 1.2.3 (Aspherical, $K(G,1)$). A complex K is said to be *aspherical* if its universal covering space is contractible. In this case, K is called an Eilenberg–Mac Lane space (or $K(G, 1)$ space) for the group $G = \pi_1(K)$.

The main result about aspherical spaces and the graph of spaces construction is the following which gives simple conditions on when the total space is aspherical. A proof can be found in [35].

Theorem 1.2.4 (Total space aspherical). *Let Γ be a graph of aspherical edge and vertex spaces with π_1-injective maps. Then the total space of this graph of spaces is also aspherical.*

Definition 1.2.5 (Graph of groups). A *graph of groups* consists of a finite graph Γ, a vertex group G_v associated to each vertex $v \in \Gamma$, an edge group G_e associated to each edge $e \in \Gamma$, and injective homomorphisms $\varphi_{\iota,e} : G_e \to G_{\iota(e)}$ and $\varphi_{\tau,e} : G_e \to G_{\tau(e)}$ for each edge e of Γ.

Definition 1.2.6 (Fundamental group of a graph of groups). Given a homomorphism $\varphi : G \to H$ between groups G and H, one can represent this by a continuous map $f : K(G, 1) \to K(H, 1)$ of Eilenberg–Mac Lane complexes. In this way one can replace a graph of groups by a graph of spaces. The total space does not depend (up to homotopy) on the choices of $K(G_v, 1)$ and $K(G_e, 1)$ spaces. The *fundamental group of the graph of groups* can be defined to be the fundamental group of the resulting graph of spaces. This point of view is developed carefully in [35].

Example. If the edge spaces are all circles, and the vertex spaces are all 2-tori, and the maps induce embeddings of the $\mathbb{Z}$ edge groups into the $\mathbb{Z}^2$ vertex groups, then the total space is aspherical. This example will be used in the next section when we conclude that *vertex spaces* are aspherical.

1.2.2 The torus construction and vertex groups

The *torus construction* is a functorial construction which takes as input a finite simplicial 2-complex K, and which produces a 2-dimensional cell complex $T(K)$ which is composed of 2-tori glued together according to the intersection pattern of the 2-simplices of K.

We define and explore elementary properties of the torus construction here. The first application of the torus construction will be in defining the vertex spaces (and vertex groups) used in the definition of the snowflake groups $G_{r,P}$. In the next chapter, we shall use functorality of the torus construction to prove upper bounds for the Dehn function of certain subgroups of 3-dimensional CAT(0) cubical groups. The torus construction is useful for producing groups with interesting geometric and topological (finiteness) properties.

Definition 1.2.7 (The Torus Construction). Let K be a finite simplicial 2-complex. The *torus complex associated to K*, denoted by $T(K)$, is the result of the following two operations:

1. First, identify all the vertices of K to one point. That is, consider the 2-dimensional cell complex $K/K^{(0)}$.

   Note that $K/K^{(0)}$ has the same number of 1-cells as K. However, now each 1-cell is a loop, and so represents a generator of $\pi_1(K/K^{(0)})$. The 2-simplices of K become length 3 relations. So $\pi_1(K/K^{(0)})$ has finite presentation; with generators in bijective correspondence with the 1-cells of K, and length 3 relations in bijective correspondence with the 2-simplices of K. Note also that $K/K^{(0)}$ is the presentation 2-complex corresponding to this presentation.

2. Second, form $T(K)$ by attaching triangular 2-cells to $K/K^{(0)}$ — one new 2-cell corresponding to each existing 2-cell of $K/K^{(0)}$ — as follows. If xyz denotes the attaching map of a 2-cell of $K/K^{(0)}$ (where x, y and z are 1-cells of $K/K^{(0)}$), then we attach a new 2-cell via the map $x^{-1}y^{-1}z^{-1}$. Here x^{-1} denotes the loop x with the opposite orientation.

Example (Properties of $T(K)$ and examples). The following properties/examples are left as exercises.

1. Property. $T(K)$ is a Δ-complex (in the sense of Hatcher's *Algebraic Topology*) with one 0-cell, the same number of 1-cells as K, and with twice as many 2-cells as K.

2. **Example.** If K consists of a single 2-simplex, then $K/K^{(0)}$ consists of a triangle with all 3 vertices identified. It is a presentation 2-complex corresponding to the following presentation of the free group of rank 2:

$$\langle\, a, b, c \mid abc \,\rangle$$

Finally $T(K)$ is the presentation 2-complex corresponding to the following finite presentation:

$$\langle\, a, b, c \mid abc,\ a^{-1}b^{-1}c^{-1} \,\rangle$$

It is easy to see that $T(K)$ is a 2-torus, with one vertex, three edges and two 2-cells. See Figure 1.2.

     In general, every 2-simplex of a finite simplicial 2-complex K will be replaced by a 2-torus (subdivided into two triangular 2-cells) in the construction of $T(K)$. This is the reason for the name *torus construction*.

     Back to the example of K consisting of a single 2-simpex. Note that the universal cover of $T(K)$ contains arbitrarily large copies of the original triangle K. For each integer $n > 0$ there is a triangle in the universal cover of $T(K)$ whose edges are subdivided into n segments, and which is tiled by n^2 2-cells.

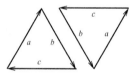

Figure 1.2: The torus construction $T(K)$ applied to a single 2-simplex.

3. **Example.** If K consists of a cone on the join of two 0-spheres (that is, K is a simplicial complex obtained from a square by subdividing as follows: connect the barycenter of the square to each of its 4 corners), then $T(K)$ has fundamental group $F_2 \times F_2$.

     Again, for each integer $n > 0$ one can find subdivided copies of the original complex K in the universal covering space of $T(K)$ (each 2-cell of K will be enlarged and tiled by n^2 2-cells in the universal covering space of $T(K)$).

4. **Example.** If K is the three-fold join of 0-spheres (that is, K is the boundary of a solid octahedron), then the fundamental group of $T(K)$ was first introduced (not in this way!) and studied by Stallings in [36]. We shall return to this example later on. Write down a presentation for this group. Visualize the universal covering space of the complex $T(K)$. Are there any large copies of K in this cover?

5. **Property.** Verify functorality of the torus construction. This will be useful in the next chapter. Let $f\colon K \to L$ be a simplicial map of finite simplicial 2-complexes K and L. Prove that there is an induced cellular map $T(f)\colon T(K) \to T(L)$ which satisfies the following two functorial properties:

   (a) $T(f \circ g) = T(f) \circ T(g)$

   (b) $T(\mathbb{I}_K) = \mathbb{I}_{T(K)}$.

   In order to define $T(f)$, determine what $T(f)$ does to the torus $T(\sigma)$ in each of the three cases where $f(\sigma)$ is a 0-simplex, a 1-simplex, and a 2-simplex.

6. **Property.** There are nice consequences obtained by combining functoriality of the torus construction with functorality of π_1. These will be used in the next chapter.

   Prove that if the finite simplicial 2-complex K is a retract of the finite simplicial 2-complex L, then the group $\pi_1(T(K))$ is a (group) retract of the group $\pi_1(T(L))$. Recall, that K a retract of L means that $i\colon K \subset L$ and that there is a simplicial map $f\colon L \to K$ such that $f \circ i = \mathbb{I}_K$.

We are now ready to define the vertex groups which are used in the graph of groups description of the snowflake groups.

Definition 1.2.8 (The vertex group V_m; geometric description). The vertex group V_m is defined as the fundamental group of the torus complex $T(K)$ of the simplicial 2-complex K obtained by taking the cone on a line segment which is composed of m 0-cells and $(m-1)$ 1-cells.

Note that K is also obtained by subdividing a $(m+1)$-gon into $m-1$ triangles, by connecting one boundary vertex to the remaining m vertices.

We choose a set of generators $\{a_1, \ldots, a_m\}$ for $\pi_1(T(K))$, and an element $c = a_1 \cdots a_m$ as follows. Orient all the 1-cells of the segment consistently (initial vertex of one cell is terminal vertex of adjacent cell), and orient the two 1-cells from the cone vertex to the endpoints of the segment so that their terminal vertices are on the segment.

Label the oriented edge from the cone vertex to the initial endpoint of the segment by a_1, and the oriented edge from the cone vertex to the terminal endpoint of the segment by c. Label the oriented 1-cells of the segment in order by $a_2, \ldots, a_m$.

Remark 1.2.9. From this geometric description it should be clear that the vertex groups V_m are 2-dimensional. The space $T(K)$ is an aspherical 2-complex. One way to see this is to show that it is homotopy equivalent to the total space of a graph of vertex 2-tori and edge circles. The underlying graph is dual to the triangulated disk K. This latter space is aspherical by Example 1.2.1.

Note that there are arbitrarily large scaled copies of the original triangulated disk in the universal cover of $T(K)$. These are seen as *scaled relations* in Figure 1.3 below.

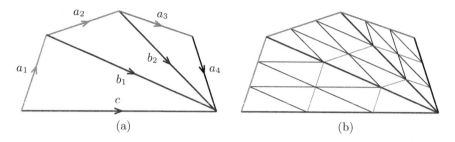

Figure 1.3: Some relations in V_4: $c = a_1a_2a_3a_4$ and $c^3 = (a_1)^3(a_2)^3(a_3)^3(a_4)^3$.

Definition 1.2.10 (Vertex groups; algebraic description). Begin with $m - 1$ copies of $\mathbb{Z} \times \mathbb{Z}$, the ith copy having generators $\{a_i, b_i\}$. The group V_m is formed by successively amalgamating these groups along infinite cyclic subgroups by adding the relations

$$b_1 = a_2b_2, \quad b_2 = a_3b_3, \quad \ldots, \quad b_{m-2} = a_{m-1}b_{m-1}.$$

Thus V_m is the fundamental group of a graph of groups whose underlying graph is a segment having $m - 2$ edges and $m - 1$ vertices. We define two new elements: $c = a_1b_1$ and $a_m = b_{m-1}$. Then $a_1, \ldots, a_m$ generate V_m and the relation $a_1 \cdots a_m = c$ holds. The element c is called the *diagonal element* of V_m.

Example (Vertex groups as right-angled Artin groups). We shall study right-angled Artin groups in the next chapter. Verify that the vertex groups V_m defined above are just right-angled Artin groups, whose defining graph is a line segment of m vertices (and $m - 1$ edges). Check also that the Artin generators of V_m are $\{c, b_1, b_2, \ldots, b_{m-1}\}$.

Remark 1.2.11 (Alternative Vertex groups). We gave a specific definition of V_m above. However, there is a lot of flexibility in defining vertex groups. An alternative version of the vertex group, V_m, could have been given as the fundamental group of a different subdivision of the $(m + 1)$-gon into $(m - 1)$ 2-simplices. This would have all the properties of the vertex group V_m defined above, and could equally well serve in the definition of the snowflake group $G_{r,P}$ which we shall describe in the next section.

Remark 1.2.12 (Snowflake groups as graphs of $\mathbb{Z}^2$). We have seen that the vertex group V_m is the fundamental group of a graph of groups with underlying graph given by the dual tree to the subdivision of the $(m + 1)$-gon into $(m - 1)$ triangles, and with all edge groups equal to $\mathbb{Z}$.

Recall that the top level description of the snowflake groups $G_{r,P}$ was as the fundamental group of a graph of groups with V_m vertex groups and $\mathbb{Z}$ edge groups. We shall see that the tree of groups decomposition of the V_m above is compatible with this graph of groups description of $G_{r,P}$.

The net result is that each of the snowflake groups is just some (large) graph of groups with $\mathbb{Z}^2$ vertex groups and $\mathbb{Z}$ edge groups.

The next three lemmas give the properties of the V_m which are analogous to the properties of $\mathbb{Z}^m$. We think of the generators a_i as standard basis elements for $\mathbb{Z}^m$, and the element c as the long diagonal of the cubical m-cell in the m-torus. These will be useful in establishing the upper bounds for the Dehn functions of the Snowflake groups. The notation in these lemmas is from the algebraic formulation in Definition 1.2.10. Proofs of these results are given in [10].

Lemma 1.2.13 (Shuffling lemma for vertex groups). *Let $w = w(a_1, \ldots, a_m, c)$ be a word representing c^N in V_m for some integer N. Let n_i be the exponent sum of a_i in w, and n_c the exponent sum of c in w. Then the words $a_1^{n_1} \cdots a_m^{n_m} c^{n_c}$ and $c^{n_c} a_m^{n_m} \cdots a_1^{n_1}$ also represent c^N in V_m and $n_i = N - n_c$ for all i.*

The corollary is easily seen to be true, e.g., by looking at Figure 1.3.

Corollary 1.2.14 (Scaling in vertex groups). *The following equations hold in V_m for each positive integer N.*

$$a_1^N \cdots a_m^N = (a_1 \cdots a_m)^N$$

This next lemma is crucial in the proof of the upper bounds for the Dehn function of the snowflake groups. It gives a more precise estimate on areas of words in the vertex groups. At first glance these groups have quadratic Dehn function, so a loop should have area bounded above by a constant multiple of the square of its length. But usually we can get better upper bounds. For example, in $\mathbb{Z}^2 = \langle a, b \mid [a, b] \rangle$ the area of the loop $[a^m, b^n]$ is bounded above by mn which is better than $(2m + 2n)^2$. In this example, only cross product terms, like mn, are important, and squares like m^2 and n^2 are not necessary for the upper bounds on the area. The lemma says that this holds true in the context of the vertex groups V_m. A proof is given in [10].

Lemma 1.2.15 (Careful area estimates in vertex groups). *Let w be a word in the generators $a_1, \ldots, a_{m-1}, b_0, \ldots, b_{m-1}$ which represents the element x^N for some N, where x is a generator a_i or b_i. Let w be expressed as $w_1 \cdots w_k$ where each w_i is a power of a generator. Then $N \leqslant |w|$ and $\mathrm{Area}(wx^{-N}) \leqslant 2 \sum_{i<j} |w_i| |w_j|$.*

1.3 Snowflake groups

In the first subsection we give two illustrative examples of snowflake groups. This motivates the formal definition of snowflake groups, and the sketch of the argument for lower bounds on their Dehn functions. In the second subsection we give a sketch of the arguments involved in the upper bounds on the Dehn functions.

1.3.1   Snowflake groups and the lower bounds

We give a formal definition of the snowflake groups $G_{r,P}$ towards the end of the subsection. We start with two concrete examples which show (1) how the log term appears in the Dehn function exponent, and (2) how the Perron–Frobenius eigenvalue of the matrix P appears in the Dehn function exponent.

Before reading these examples, think about how to construct a space from 2-tori and cylinders whose fundamental group has a Dehn function of the form $x^{2\log_2(3)}$. Is it even possible? If so, how many 2-tori and cylinders are needed? How many 2-tori and cylinders would be used to construct a group with Dehn function of the form $x^{p/q}$ for a rational number $p/q \in [2.\infty)$? What about using finitely many 2-tori and cylinders to construct a space whose fundamental group has Dehn function of the form $x^{2\log_\lambda(5)}$, where λ is the Perron–Frobenius eigenvalue of the matrix $P = \left(\begin{smallmatrix} 1 & 1 \\ 2 & 1 \end{smallmatrix}\right)$?

The first example; competing exponential growth rates.

The following group demonstrates the key features of the snowflake group complexes; namely aspherical spaces, whose universal covers contain embedded diagrams which have a fractal (snowflake) nature. The geometry of these diagrams depend on growth rates of strips on the one hand, and the growth rate of a dual tree on the other hand. It is the comparison of these two exponential growth rates that gives the log term in the exponent of the Dehn function.

$$G \;=\; \langle a_1, a_2, c, s_1, s_2 \,|\, a_1 a_2 = c = a_2 a_1,\, s_i^{-1} a_i^r s_i = c \rangle$$

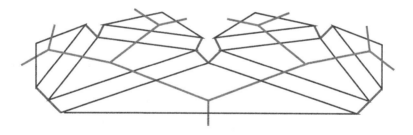

Figure 1.4: Half of snowflake diagram and dual tree.

Note that G is a graph of groups with underlying graph a bouquet of two circles. The vertex group is $\mathbb{Z}^2$ (generated by $\{a_1, a_2, c\}$) and the edge groups are both $\mathbb{Z}$. The stable letters conjugate an rth power of a generator a_i to the element c. Assume r is a positive integer for now. The general case is treated in detail in [10]. The total space of the corresponding graph of spaces is just a 2-torus with two cylinders attached. The universal cover of this space consists of a collection of planes (covers of the torus) connected together in a tree like fashion by strips (covers of the cylinders). The tree which describes the way in which planes are attached together via strips is just the Bass-Serre tree of this graph of spaces. This universal cover is a contractible 2-complex.

We construct an embedded disk with large area (as a function of boundary length) in the universal cover inductively as follows. Given a positive integer k, start with a c-line segment of length r^k in some plane. We can write $c^{r^k} = a_1^{r^k} a_2^{r^k}$; this is represented by the base triangle in Figure 1.4. Note that we can also write $c^{r^k} = a_2^{r^k} a_1^{r^k}$. This would be represented by a second triangle based on c^{r^k}. This will develop into the "lower half" of the snowflake diagram. Schematically (and geometrically) it will look like a reflected image of the top half of the diagram which is shown in Figure 1.4. Only the labels on edges will be different. So, we can just restrict attention to the upper half of the diagram.

There are strips attached to the base plane along a_i lines. There is an s_i strip attached to the $a_i^{r^k}$ side of the base triangle. The other edge of these strips are represented by $c^{r^{k-1}}$ in new planes. Thus, we have a short path,

$$s_1 c^{r^{k-1}} s_1^{-1} s_2 c^{r^{k-1}} s_2^{-1}$$

which has the same endpoints as c^{r^k}. The length of this path is $4 + 2r^{k-1}$ which is less than r^k for large r (greater than 2). We have approximately divided the length of the original path by a factor of $r/2$, at the expense of traversing strips. We repeat this shortening procedure; now applying the same trick to each of the two $c^{r^{k-1}}$ subsegments. After k iterations, the c subsegments will have length 1, and we stop the shortening procedure. The resulting path is the outer boundary along the top of the diagram in Figure 1.4. As a word in the a_i, s_i and c, this boundary word is obtained from c^{r^k} by iteratively replacing every maximal c-segment, which will be a c^{r^m}, by the expression $s_1 c^{r^{m-1}} s_1^{-1} s_2 c^{r^{m-1}} s_2^{-1}$. The values of m will decrease at each stage, from k down to 1. Note that the top boundary of the diagram can be collapsed in a 2-to-1 fashion onto the edges of the dual tree. Thus the length of the boundary grows as the dual tree grows; that is, as 2^k.

The whole snowflake diagram is obtained by "reflecting" the half shown in Figure 1.4 over the horizontal base line. We call this the "diameter" of the diagram. It is just c^{r^k} and so has length r^k.

Finally, the area of the diagram is at least the area of the triangular regions adjacent to the "diameter". This is at least $(r^k)^2$. We summarize this discussion as follows. The reader should fill in details as an exercise.

- Length of "diameter" grows as r^k.

- Area is at least square of "diameter".

- Length of boundary, ∂_k, grows as dual tree (2^k). Specifically, there are constants $C < D$ independent of k, so that

$$C2^k \leqslant |\partial_k| \leqslant D2^k.$$

- Area $\geqslant (r^k)^2 = (2^k)^{2\log_2(r)} \geqslant D^{-2\log_2(r)}|\partial_k|^{2\log_2(r)}$.

What we have is an sequence of embedded diagrams (one for each positive integer k) in a contractible 2-complex whose area is at least a constant times the boundary length to the power $2\log_2(r)$. There are no more efficient ways of filling these boundary curves. Any other filling will have to agree homologically with the embedded disks (otherwise one would obtain a nontrivial 2-cycle in a contractible 2-complex, a contradiction). In particular, any other filling of the boundary loops has to contain at least each of the 2-cells in the embedded disk fillings. Thus we have found a sequence of exponentially growing loops in the universal cover with area bounded below by length to the power $2\log_2(r)$. Use the definition of equivalence of functions to show that this sequence of loops is enough to conclude that the Dehn function δ_G satisfies

$$x^{2\log_2(r)} \preceq \delta_G(x).$$

See Remark 1.1 of [9] for details.

We summarize the key components in the Dehn function exponent, $2\log_2(r)$.

- The initial factor of 2 corresponds to the fact that area is a quadratic function of length in the Euclidean plane (and this fact about areas will remain the same for more general vertex groups V_m)

- The base of the logarithm dependes on the growth rate of the dual tree. We shall see in the next example, that this can be made to correspond to the growth rate of an irreducible matrix P.

- The input to the logarithm is the number r, which depends on how the cylinders are attached to the vertex spaces in the construction of the snowflake groups.

Remark 1.3.1 (Rational exponents). Note if we want to see that the rational numbers in $[2,\infty)$ are in $\mathbb{P}$, we just need to replace the 2-torus in the example above by the vertex group $V_{2^{2q}}$, and use the value 2^p for r. That is, we start with a collection of $2^{2q} - 1$ 2-tori glued together as in the definition of $V_{2^{2q}}$. There are generators a_i for $1 \leqslant i \leqslant 2^{2q}$, and there is a diagonal element c. Now attach 2^{2q} cylinders to this collection of 2-tori, so that the ith cylinder attaches to c at one end and to $a_i^{2^p}$ at the other end.

By arguing in a similar fashion as above, we conclude that there is a lower bound of $x^{2\log_{2^{2q}}(2^p)} = x^{p/q}$ for the Dehn function of this group. The upper bounds will be established in the next section.

The second example; Perron–Frobenius eigenvalue of the matrix P.

Here we are given the following data: a matrix $P = \left(\begin{smallmatrix} 1 & 1 \\ 2 & 1 \end{smallmatrix}\right)$ and a rational number $r > 3$ (take r to be an integer for simplicity). This determines a graph of spaces (graph of groups) as follows.

- P is the transition matrix for a graph Γ. Figure 1.5 shows the directed graph where the vertices have been blown up into polygons (a triangle and a trapezoid). Γ is the underlying graph in a graph of groups description of $G_{r,P}$.

- Edge groups are all $\mathbb{Z}$. Vertex groups are V_{m_j}, where m_j is jth row sum of P. So, the triangle in Figure 1.5 represents the fact that V_1 has 2 generators and a diagonal element, c_1. The trapezoid represents the fact that V_2 has 3 generators and a diagonal element, c_2.

- As before, the stable letters s_i conjugates an rth power of the generator a_i to a diagonal element c. The only thing to note is that we use the matrix P and the graph Γ to specify which vertex group the target diagonal element should lie in. Figure 1.5 shows this for the current example. A description of the general situation is given after this example.

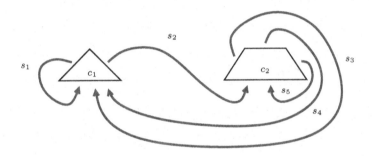

Figure 1.5: Schematic diagram of the 2-complex for the group $G_{r,P}$.

In Figure 1.6 we see a typical snowflake diagram in for the group $G_{r,P}$ of this example. (Actually, this disk starts with a power of a diagonal element c which may not be a power of r, so there are small error terms in attaching strips at each stage. One can see these at the one end of each strip in this figure.)

Looking back on the three key points in the summary statement of the last example, we note that the initial factor of 2 remains, the input r will remain the same. The only difference is the growth rate of the dual tree. One can see that vertices of the dual tree which are inside of triangles have valence 3, while vertices of the dual tree which are inside of trapezoids have valence 4. So, it is not a uniform tree.

It is a nice exercise to see that this dual tree grows as powers of the transpose of the matrix P, and hence as powers of the Perron–Frobenious eigenvalue λ. (Hint. It is easiest to view the snowflake diagram (or equivalently the dual tree) as being build up in layers. The number of triangles on layer $m + 1$ is equal to the sum of the number of triangles on layer m and 2 times the number of trapezoids on layer m. Similarly, the number of trapezoids on layer $m + 1$ equals the sum of the number of triangles on layer m and the number of trapezoids on layer m.) Arguing as in the previous example, we conclude that $x^{2\log_\lambda(r)}$ is a lower bound for the Dehn function of this group.

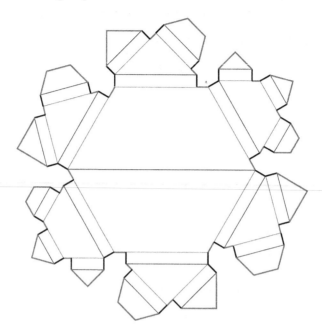

Figure 1.6: A snowflake disk based on the matrix $P = \left(\begin{smallmatrix} 1 & 1 \\ 2 & 1 \end{smallmatrix}\right)$.

Now we give the formal definition of the snowflake groups corresponding to an arbitrary irreducible integer matrix P with leading eigenvalue greater than 1, and to a rational number r greater than the largest row sum of P.

Definition 1.3.2 (Snowflake groups $G_{r,P}$). Start with a non-negative square integer matrix $P = (p_{ij})$ with R rows. Let m_i be the sum of the entries in the ith row and let $n = \sum_i m_i$, the sum of all entries. Form a directed graph Γ with vertices $\{v_1, \ldots, v_R\}$ and having p_{ij} directed edges from v_i to v_j. Label the edges as $\{e_1, \ldots, e_n\}$ and define two functions $\rho, \sigma \colon \{1, \ldots, n\} \to \{1, \ldots, R\}$ indicating the initial and terminal vertices of the edges, so that e_i is a directed edge from $v_{\rho(i)}$ to $v_{\sigma(i)}$ for each i. These functions also indicate the row and column of the matrix entry accounting for e_i. Partition the set $\{1, \ldots, n\}$ as $\bigcup_i I_i$ by setting $I_i = \rho^{-1}(i)$. Note that $|I_i| = m_i$.

Choose a rational number $r = p/q$ with $(p, q) = 1$. We define a graph of groups with underlying graph Γ as follows. The vertex group G_{v_i} at v_i will be V_{m_i} and all edge groups will be infinite cyclic. Relabel the standard generators of these vertex groups as $\{a_1, \ldots, a_n\}$ in such a way that the standard generating set for G_{v_i} is $\{a_j \mid j \in I_i\}$. Let c_i be the diagonal element of the vertex group

G_{v_i}. Then the inclusion maps are defined by mapping the generator of the infinite cyclic group G_{e_i} to the elements $a_i{}^p \in G_{v_{\rho(i)}}$ and $c_{\sigma(i)}{}^q \in G_{v_{\sigma(i)}}$.

Let s_i be the stable letter associated to the edge e_i. The full path group of this graph of groups has presentation

$$\langle G_{v_1}, \ldots, G_{v_R}, s_1, \ldots, s_n \mid s_i^{-1} a_i{}^p s_i = c_{\sigma(i)}{}^q \text{ for all } i \rangle.$$

The fundamental group $G_{r,P}$ of the graph of groups is obtained by adding relations $s_i = 1$ for each edge e_i in a maximal tree in Γ. However, we shall continue to use the generating set $\{a_1, \ldots, a_n, s_1, \ldots, s_n\}$ for $G_{r,P}$ even though some of these generators are trivial.

The definition of *snowflake words* below requires that r be greater than the maximum row sum of P. These snowflake words are used to produce the boundary of an embedded Van Kampen diagram in the universal cover of the presentation 2-complex for $G_{r,P}$ in the proof of the lower bounds. Also, a key step in the proof of the upper bounds involves proving that snowflake words are close to being geodesics (their length is a definite linear function of geodesic length).

Definition 1.3.3 (Snowflake words). For a fixed integer $N_0 > 0$ we define snowflake words recursively as follows. A word w representing c^N for some diagonal element c of a vertex group is a *positive snowflake word* if either

(i) $|w| \leqslant |N| \leqslant N_0$, or

(ii) $w = (s_{i_1} u_1 s_{i_1}^{-1})(a_{i_1}^{N_1}) \cdots (s_{i_m} u_m s_{i_m}^{-1})(a_{i_m}^{N_m})$ where each u_j is a positive snowflake word representing a power of $c_{\sigma(i_j)}$ and $|N_j| < p$ for all j. Here $\{a_{i_1}, \ldots, a_{i_m}\}$ is the standard ordered generating set for the vertex group containing c as its diagonal element.

In the second case note that each subword $(s_{i_j} u_j s_{i_j}^{-1})(a_{i_j}^{N_j})$ represents a power of a_{i_j}, and by Lemma 1.2.13 this power is N. Then since $|N_j| < p$, the word $(s_{i_j} u_j s_{i_j}^{-1})$ represents either $a_{i_j}{}^{\lfloor N/p \rfloor p}$ or $a_{i_j}{}^{\lceil N/p \rceil p}$. Consequently, the word u_j represents either $c_{\sigma(i_j)}{}^{\lfloor N/p \rfloor q}$ or $c_{\sigma(i_j)}{}^{\lceil N/p \rceil q}$.

A *negative snowflake word* is defined similarly, except that the ordering of the terms representing powers of a_{i_j} is reversed. This is achieved by replacing condition (ii) with

(ii′) $w = (s_{i_m} u_m s_{i_m}^{-1})(a_{i_m}^{N_m}) \cdots (s_{i_1} u_1 s_{i_1}^{-1})(a_{i_1}^{N_1})$ where u_j is a negative snowflake word representing a power of $c_{\sigma(i_j)}$ and $|N_j| < p$ for all j.

As with positive snowflake words, each word u_j will represent either $c_{\sigma(i_j)}{}^{\lfloor N/p \rfloor q}$ or $c_{\sigma(i_j)}{}^{\lceil N/p \rceil q}$.

Remark 1.3.4. N_0 will be chosen so that the snowflake process will actually shorten c^N whenever $N \geqslant N_0$.

Proposition 1.3.5. *Given $G_{r,P}$ there are positive constants A_0, A_1 with the following property. If c is the diagonal element of one of the vertex groups and w is a snowflake word representing c^N, then $A_0 |w|^\alpha \leqslant N \leqslant A_1 |w|^\alpha$, where $\alpha = \log_\lambda(r)$.*

The lower bounds. Now we have the lower bounds for the Dehn function of $G_{r,P}$. Complete the following steps.

1. Let w_+ (resp. w_-) be positive (resp. negative) snowflake words for c^N. Then the area of the word $w = w_+ w_-^{-1}$ is at least N^2, which is at least of order $|w|^{2\alpha}$ by Proposition 1.3.5.

2. The Van Kampen diagram for $w_+ w_-^{-1}$ *embeds* into the contractible universal cover of the presentation 2-complex of $G_{r,P}$.

3. Any filling of the loop $w_+ w_-^{-1}$ will use at least as many 2-cells as the embedded Van Kampen diagram above.

Example. Run the lower bound arguments explicitly in the following cases.

1. P is the (1×1)-matrix (2) and $r = 3$.

2. $P = \begin{pmatrix} 1 & 2 \\ 1 & 1 \end{pmatrix}$ and $r = 4$.

1.3.2   Upper bounds

In order to prove the upper bound one has to consider an arbitrary word, w, in the generators of $G_{r,P}$ which represents the identity element. One can cut a Van Kampen diagram for such a word, w, along s_i-corridors.

It turns out to be easier to give an inductive proof of upper bounds for the two haves of the cut diagram. This is given in Proposition 1.3.9 below. In order to perform the inductive step, we need two other ingredients. The first is the area bound in the vertex groups, given in Lemma 1.2.15. Secondly, we need to estimate the lengths of the s_i corridors as a function of the length along the boundary of the word w. An upper bound for this is given by the distortion of the c_j-subgroups in the ambient group. This is the content of Corollary 1.3.8. Here are some more details

The approach to understanding distortion of the c_j-subgroups in $G_{r,P}$ is to follow the most naive expectations. Suppose one wishes to estimate the distance in $G_{r,P}$ between the endpoints of some large power c_j^N of a diagonal element c_j in a vertex group V_{m_j}. The first naive step is to write c_j^N as a product of Nth powers of the generators of V_{m_j}. Then replace each a_i^N by an s_i-conjugate of an N/r-th power of some other c_k and repeat the process. This terminates in the snowflake representative w_{sf} of c_j^N. We know from Proposition 1.3.5 that N is bounded above by a constant (depending only on P and r) times $|w_{\mathrm{sf}}|^{\log_\lambda(r)}$. What we have to show is that N is bounded above by some constant (depending only on P and r) times $|w_{\mathrm{geo}}|^{\log_\lambda(r)}$, where w_{geo} denotes a geodesic between the endpoints of c_j^N.

This is done by proving that the lengths of geodesic and snowflake words having the same endpoints are linearly related. We see this by inductively comparing lengths of geodesics and snowflake words. To make the induction work we argue that geodesics behave in the same "locally greedy" way that snowflake words do; namely, they cross over from the expanding (or c) side of an s_i-strip to the contracting (or a_i) side of the strip in order to shorten lengths. Specifically, a geodesic path whose endpoints both lie on the contracting side of an s_i-strip will never cross over that strip. The proof makes essential use of the hypothesis that r is larger than the maximum row sum of P. See [10] for details.

Lemma 1.3.6 (Geodesics never cross strips in expanding direction). *Let w be a geodesic in $G_{r,P}$ representing an element of a vertex group V_m. Then w is a product of subwords $w_1 \cdots w_k$ where each w_i is a power of a generator a_j, or begins with s_j and ends with s_j^{-1} for some j.*

The preceding lemma is a key ingredient in the inductive proof of the next proposition, which compares the lengths of geodesic paths and snowflake paths with the same endpoints. Again details can be found in [10].

Proposition 1.3.7 (Geodesics and snowflake words linearly related). *Let $\alpha = \log_\lambda(r)$ where λ is the Perron–Frobenius eigenvalue of P. Given $G_{r.P}$ there is a constant $B \geqslant 1$ with the following property. If c is the diagonal element of one of the vertex groups and c^N (with $N \geqslant 0$) is represented by both a snowflake word w_{sf} and a geodesic w_{geo}, then $|w_{sf}| \leqslant B |w_{geo}|$.*

Since Proposition 1.3.5 has already estimated lengths of snowflake words, we can combine it with the previous result to obtain an upper bound of the distortion of the infinite cyclic $\langle c_j \rangle$ subgroups of $G_{r,P}$. This is precisely the estimates on the lengths of s_j-corridors that we want.

Corollary 1.3.8 (Distortion of $\langle c_j \rangle$ subgroups). *Given $G_{r,P}$ there is a constant $C \geqslant 1$ with the following property. If w is a word representing c^N for some N, where c is the diagonal element of one of the vertex groups V_m, then $N \leqslant C |w|^\alpha$.*

Now we have sufficient geometric control on the lengths of s_i-corridors. We combine this with the careful area estimates of Lemma 1.2.15 to give an inductive proof of the area upper bounds. Note that on taking $N = 0$ we obtain the desired upper bounds.

Proposition 1.3.9 (Area upper bounds). *Let $\alpha = \log_\lambda(r)$ where λ is the Perron–Frobenius eigenvalue of P. Let w be a word in $G_{r,P}$ representing x^N for some N, where x is either a generator a_i or the diagonal element of one of the vertex groups V_m. Then $\mathrm{Area}(wx^{-N}) \leqslant r^2 C^2 |w|^{2\alpha}$ (with C given by Corollary 1.3.8).*

Here is a sketch of the proof. There is enough detail to see why the distortion of the c_j-subgroups and the careful area estimates are the key ingredients in establishing upper bounds. It is a "generic situation" proof, where there are all three type of subwords as described below. It gives the flavor of how the careful estimates fit nicely with some simple algebra to give an inductive proof. A very careful proof is given in [10].

We note that the word w represents an element x^N of a vertex group. As an edge path in the Cayley complex, w will start and end in a coset of the vertex group. If it leaves the coset along an s_i-strip, it must return at some point. The corresponding first-return subword of w will be of the form $s_i w' s_i^{-1}$, and represents a power of c in the vertex group. Similarly, if it leaves the coset along an s_i^{-1}-strip it must also eventually return. In this case, the corresponding first-return subword of w will be of the form $s_i^{-1} w'' s_i$, and represents a power of a_i in the vertex group. These first return subwords partition w into a collection of subwords w_j. There are three types of subword; those that never leave the coset, those first-return words of the form $s_i^{-1} w'' s_i$, and those first-return subwords of the form $s_i w' s_i^{-1}$. Let l_j denote the length of w_j, so that $|w| = \sum_j l_j$.

If w_j is of the form $s_i w' s_i^{-1}$ or $s_i^{-1} w'' s_i$, then it represents a power of an element $x_j^{N_j}$ in the vertex group, and we know by induction on length that

$$A_j = \text{Area}(w_j x_j^{-N_j}) \leqslant r^2 C^2 l_j^{2\alpha} = r^2 C^2 (l_j^\alpha)^2. \tag{$*$}$$

Denote this power of x_j by m_j. We know from Corollary 1.3.8 that $m_j \leqslant C l_j^\alpha$.

Now we have expressed x^N as a product of words which are entirely contained in the vertex group; if a subword w_j was not contained in the vertex group, then it was one of the two described in the previous paragraph, and was replaced by an m_j power of some x_j in the vertex group. Define new subwords w_j' by, $w_j' = w_j$ if w_j is already contained in the vertex coset, and w_j' is the appropriate power of x_j otherwise. By the careful area estimates of Lemma 1.2.15, we conclude that the area of this expression is

$$A_0 \leqslant 2 \sum_{i<j} |w_i'||w_j'| \leqslant 2 \sum_{i<j} (r C l_i^\alpha)(r C l_j^\alpha) \leqslant r^2 C^2 2 \sum_{i<j} l_i^\alpha l_j^\alpha.$$

The estimate $|w_j'| \leqslant r C l_i^\alpha$ comes from considering the worst case, namely when w_j is of the form $s_i^{-1} w'' s_i$, and using Corollary 1.3.8.

Combining this area with the area estimates in $(*)$ gives an upper bound for the original area:

$$\text{Area} = A_0 + \sum_j A_j \leqslant r^2 C^2 \left(\sum_j (l_j^\alpha)^2 + 2 \sum_{i<j} l_i^\alpha l_j^\alpha \right).$$

The last term above is just an expanded square. So we get

$$\text{Area} \leqslant r^2 C^2 \left(\sum_j l_j^\alpha \right)^2 \leqslant r^2 C^2 \left(\left(\sum_j l_j \right)^\alpha \right)^2 = r^2 C^2 l^{2\alpha} = r^2 C^2 |w|^{2\alpha}$$

and the inductive step is completed.

1.4 Questions and further explorations

There are lots of interesting questions and directions in which to develop the ideas in this chapter. We list some specific questions regarding Dehn functions and the torus construction, and variations on the snowflake construction below.

Question 1.4.1 (The torus construction and Dehn functions). We shall see in the next chapter, that certain combinatorial properties of a 2-complex K give rise to interesting lower bounds for the Dehn function of the fundamental group of the resulting torus complex, $T(K)$.

1. There are lots of open problems concerning Dehn functions of groups resulting from the torus construction. For example, if one starts with the 3-fold join of 0-spheres, then the torus construction yields a 2-complex whose fundamental group was introduced by Stallings in [36]. What is the Dehn function of Stallings' group? It is known (private communication of Tim Riley and Murray Elder) that it's Dehn function is bounded below by x^2 and above by $x^{5/2}$, but an exact determination of the Dehn function has yet to be found.

2. Are there interesting variations on the torus construction? Can one start with a simplicial 2-complex and replace each 2-cell by a suitably triangulated one vertex high genus surface (instead of a 2-torus)? Perhaps one should replace certain finite collections of 2-cells (in some configuration) by suitably triangulated high genus surfaces. The previous sentences are a little vague, but the reader should consider the kernel of the branched cover example (final example in the next chapter) as a guiding/motivating example. What Dehn functions would the fundamental groups of the resulting complexes have? Would these fundamental groups contain $\mathbb{Z}^2$ subgroups?

3. One can run versions of the torus construction in higher dimensions. For example, two tetrahedra and a solid octahedron tile a 3-torus, so one could define a torus construction for 3-complexes. The level sets of Morse functions on high dimensional right-angled Artin groups (see later) give a good source of examples on which to base high dimensional versions of the torus construction.

Question 1.4.2 (Dehn functions and subdivision rules). There are other ways of obtaining diagrams with large area. For example, by iterating the subdivision rule shown in Figure 1.7 k times, one obtains a diagram with boundary length $5(2^k)$ and area 6^k. As k tends to infinity, these diagrams give an isoperimetric estimate of $x^{\log_2(6)}$ which is more than quadratic. Can any of this be modeled using Cayley graphs of groups? If so, do any of the resulting groups not contain $\mathbb{Z}^2$ subgroups?

The question about $\mathbb{Z}^2$ subgroups above is related to some very interesting questions. With just one exception (which we'll see at the end of the next chapter), every group whose Dehn function is known to be at least quadratic contains a $\mathbb{Z}^2$ subgroup or a Baumslag-Solitar $BS(m, n)$ subgroup. In fact the way that large area van Kampen diagrams are built up (at least in the sub-exponential case) seems to be by gluing pieces of Euclidean planes together. This is certainly true of the snowflake groups here, and is also true of the S-machine groups in the work of Briget-Rips-Sapir. Large portions of Euclidean planes can be seen in fillings of maximal loops in nilpotent groups in Tim Riley's section.

There is a version of the hyperbolization conjecture of 3-manifolds in group theory. It asks if every group which has a finite $K(G, 1)$ space, and has no Baumslag-Solitar or $\mathbb{Z}^2$ subgroups must be hyperbolic. If this were true, it would mean that one needs either large portions of the Baumslag-Solitar complexes or of the Euclidean plane in building up large area van Kampen diagrams in groups with finite $K(G, 1)$ spaces.

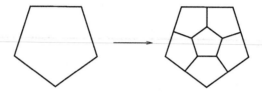

Figure 1.7: A pentagonal subdivision rule.

Question 1.4.3 (Dehn functions and distortion). Note that the $\mathbb{Z}$ edge groups are all highly distorted in the $G_{r,P}$. In fact, the distortion of the edge groups in $G_{r,P}$ is given by $f(x) \sim x^{\log_\lambda(r)}$, and it is this distortion that gives the Dehn function of $\delta(x) = (f(x))^2$.

There are variations on this construction where one uses higher rank free groups as edge groups instead of $\mathbb{Z}$ as in the snowflake groups above. This will increase the range of exponents, since one can replace the numerator and denominator of the rational number r by growth rates of monomorphisms (automorphisms) of free groups. Are all $\mathbb{Z}$ subgroups of these groups undistorted? Do these types of examples embed into non-positively curved groups such as the CAT(0) groups introduced in the next chapter?

Question 1.4.4 (Higher dimensional versions of Dehn functions). Let r be a positive integer. The snowflake groups $G_{r,P}$ admit monomorphisms ϕ_r which take the generators of the vertex groups to their rth powers, and which fix all the stable letters s_i. The *suspended snowflake group* $\Sigma G_{r,P}$ is defined to be the multiple (2-fold) ascending HNN extension of $G_{r,P}$ where each of the stable letters acts via ϕ_r. It is the fundamental group of a graph of groups with one vertex group and two edge groups. The vertex and edge groups are all $G_{r,P}$, the two monomorphisms from the each edge group are the identity and ϕ_r. It can be shown that the total space of the corresponding graph of spaces is an aspherical 3-dimensional classifying space for $\Sigma G_{r,P}$. Furthermore, in [10] it is shown that the second order Dehn function (measure of the complexity of filling 2-spheres with 3-balls) of $\Sigma G_{r,P}$ is the same as the ordinary Dehn function of $G_{r,P}$. This suspension procedure can be iterated, yielding groups $\Sigma^k G_{r,P}$ with aspherical, $(k+2)$-dimensional classifying spaces, whose order $k+1$ Dehn functions (filling $(k+1)$-spheres with $(k+2)$-balls) are identical to the Dehn function of $G_{r,P}$. The arguments and constructions in [10] rely heavily on the scalability of the vertex groups, and on the existence of the monomorphisms ϕ_r at every stage.

Are there other ways of producing high dimensional groups which display a range of higher order Dehn function behavior? In particular, are there versions of the work of Birget–Rips–Ol'shanskii–Sapir ([34] and [6]) or Sapir–Ol'shanskii for higher order Dehn functions?

Chapter 2

Dehn Functions of Subgroups of CAT(0) Groups

In this chapter we give a brief review of some notions of non-positive curvature in geometric group theory. We say that a geodesic metric space is non-positively curved if geodesic triangles are at least as thin as triangles in the Euclidean plane. Non-positively curved spaces in this sense are called CAT(0) spaces, and groups which act properly discontinuously and cocompactly by isometries on a CAT(0) space are called CAT(0) groups.

One nice property of CAT(0) groups is that there is a quadratic upper bound for their Dehn functions. This follows from the non-positive curvature of the underlying CAT(0) spaces on which the CAT(0) groups act. A detailed statement and proof appears as Theorem 6.2.1 of [15].

Here are motivating questions for the examples and constructions in this chapter. *What are possible ranges of Dehn functions for subgroups of* CAT(0) *groups? Are there gaps in the corresponding* $\mathbb{P}$ *(other than 1 to 2)? If so, do these have any special geometric or algebraic significance?* After all, the only in the usual $\mathbb{P}$ spectrum distinguishes between hyperbolic and non-hyperbolic groups, and corresponds to the sub-quadratic characterization of hyperbolicity. Is there something analogous for subgroups of non-positively curved groups? Are there other gaps?

What types of problems does one encounter when analyzing subgroups of non-positively curved groups? Are they not all non-positively curved as well? To get some feeling for the issues involved, let's ask the same question about subspaces of non-positively curved spaces.

Take Euclidean 3-space as our ambient non-positively curved space. Consider the following two subspaces; a flat plane and a 2-sphere. The plane is non-positively curved in its own metric, and the metric on it is the same as the restriction of the ambient metric. It is not *distorted* in Euclidean space. The intrinsic metric on the 2-sphere however is positively curved (two great circles from the north pole will reconverge at the south pole, this is a positive curvature phenomenon). Also, the distance between points on the sphere (as measured on the sphere), is greater than the ambient Euclidean space distance. The sphere is said to be *distorted* in the ambient space.

More interesting examples occur with negative curvature. Take a point p in hyperbolic 3-space, and an infinite geodesic ray γ based at p. For $t > 0$ let S_t denote the sphere in hyperbolic space of radius t centered at the unique point on γ which is distance t from p. As $t \to \infty$ these spheres converge (uniformly on compact sets) to a space S_∞ which contains p. This is called the *horosphere* through p, based at the point at infinity determined by γ. It turns out that S_∞ is intrinsically a Euclidean plane (not negatively curved like the ambient space). Moreover, it is exponentially distorted in the ambient hyperbolic space; in fact two points of the horosphere which are distance x apart in the ambient hyperbolic metric, will be on the order of e^x apart in the intrinsic metric on the horosphere.

For variable non-positive curvature we get more exotic horospheres. For example, there are horospheres in the product of two hyperbolic planes which are metrically 3-dimensional geometry *Sol*, and horospheres in complex hyperbolic space are metrically the 3-dimensional geometry *Nil*.

It is interesting to ask to what extent this type of behavior is mirrored by finitely presented subgroups of CAT(0) groups. Are there subgroups whose Cayley graph is so distorted or folded back and forth in the ambient Cayley graph that their intrinsic geometry (as measured by the Dehn function for example) is far from CAT(0)? The examples that follow below give an indication of what can happen with the geometry of subgroups of CAT(0) groups.

We present three families of CAT(0) groups which contain subgroups with large Dehn functions. The first family are products of free groups with free-by-cyclic groups and are described in Section 2.3. The second family are right-angled Artin groups and are described in Section 2.5. The last is a hyperbolic group which is described in Section 2.6.

This chapter is organized as follows. In Section 2.1 we recall the definitions and main results about CAT(0) groups, we consider M_κ-complexes and the link condition for non-positive curvature, and pay special attention to the case of cubical complexes and the flag link condition. In Section 2.2 we introduce Morse functions on affine complexes, and give a local Morse criterion for recognizing free-by-cyclic groups. This criterion provides a gentle introduction to the use of Morse functions, by restricting attention to 2-complexes. This also provides the background for the description of the free-by-cyclic examples in Section 2.3. In Section 2.4 we revisit the notion of Morse functions on complexes, and study the geometry and topology of kernel subgroups of CAT(0) cubical groups. Here the cubical complexes range from 1- to 3-dimensional. The Morse theory developed in Section 2.4 is important for a description of the kernels of the right-angled Artin family (provided in Section 2.5) and of the kernel of the final hyperbolic example (provided in Section 2.6).

It is important to keep in mind when reading through this chapter, that all the examples are carefully constructed. In general it appears to be very hard to determine the geometry of a distorted subgroup of a CAT(0) group. There could be a wealth of brand new geometries out there among these subgroups. Keep searching.

2.1 CAT(0) spaces and CAT(0) groups

In the first subsection we recall the basic definitions and properties of CAT(0) spaces and groups. In the second subsection we describe how to build locally CAT(0) spaces out of convex cells from the model spaces. The *link condition* is given. This is a condition for recognizing when such a space is locally CAT(0). In the final subsection, we restrict attention to the special case of piecewise Euclidean cubical complexes and the flag condition.

2.1.1 Definitions and properties

We begin with a quick review of terminology of CAT(0) spaces and CAT(0) groups. We will just give statements of main results. Other treatments are available in [14], [17], and in [21]. Many detailed proofs are given in [17].

A metric space X with metric d is said to be a *geodesic metric space* if for all $x, y \in X$ there exists a rectifiable path γ from x to y whose length is equal to $d(x, y)$. Let X be a geodesic metric space, and let $x, y, z \in X$. A *geodesic triangle in X* determined by x, y, z is simply a union $\Delta(x, y, z)$ of geodesic paths $[x, y]$, $[y, z]$ and $[z, x]$. The geodesic triangle is not necessarily unique, since geodesics are not a priori unique.

The CAT(κ) definitions involve comparing triangles in a given geodesic metric space with triangles in some model space of constant curvature. The model spaces are the 2-dimensional geometries of constant curvature κ and are denoted by M_κ^2. Special cases include M_0^2 which is the Euclidean plane, M_1^2 which is the round 2-sphere of radius 1, and M_{-1}^2 is the hyperbolic plane.

Definition 2.1.1 (Comparison triangles, comparison points). A *comparison triangle* for a triangle $\Delta(x, y, z)$ in a geodesic metric space is a triangle $\Delta(x', y', z')$ in a model space M_κ^2 so that corresponding edges are isometric. Comparison triangles are unique up to isometry.

Let $p \in [x, y]$ be a point on a geodesic triangle $\Delta(x, y, z)$. Let $\Delta(x', y', z')$ be a comparison triangle for $\Delta(x, y, z)$ in the model space M_κ^2. A *comparison point for p* is a point $p' \in [x', y']$ such that $d(x, p) = d'(x', p')$.

Definition 2.1.2 (CAT(κ) inequality, CAT(κ) spaces). A geodesic triangle $\Delta(x, y, z)$ is said to satisfy a CAT(κ) inequality if for all p on the interior of one edge, the distance from p to the opposite vertex in X is bounded above by the distance between the comparison point and corresponding opposite vertex in the model space M_κ^2. In the case that $\kappa > 0$ we require that the perimeter of $\Delta(x, y, z)$ be less than twice the diameter of the model sphere M_κ^2. In the case $\kappa \leqslant 0$ there is no restriction on the perimeter of triangles.

A geodesic space in which triangles satisfy the CAT(κ) inequality is called a CAT(κ) *metric space*.

Remark 2.1.3. One can talk about locally CAT(κ) spaces. If $\kappa \leqslant 0$, then a simply connected locally CAT(κ) space is globally CAT(κ). Locally CAT(κ) spaces for $\kappa \leqslant 0$ are called *non-positively curved* spaces.

An important fact about CAT(0) spaces is that they are contractible. The CAT(0) condition is sufficient to show that one can retract the space radially along geodesic rays to any given base point.

Theorem 2.1.4 (CAT(0) implies contractible). *If X is a CAT(0) space, then X is contractible.*

Now CAT(0) groups are defined as CAT(0) isometry analogues of uniform lattices in Lie groups.

Definition 2.1.5 (CAT(κ) groups). Let $\kappa \leqslant 0$. A group G is a CAT(κ) *group* if it acts properly discontinuously and cocompactly by isometries on a CAT(κ) space.

The next two results relate the notion of CAT(κ) groups for $\kappa \leqslant 0$ to the other notion of negatively curved groups that we have seen in this course; namely, hyperbolic groups.

Theorem 2.1.6 (CAT(-1) implies hyperbolic). *If G is a CAT(-1) group, then G is hyperbolic.*

Theorem 2.1.7 (CAT(0) and no flat planes implies hyperbolic). *If G is a CAT(0) group, and the CAT(0) space on which G acts properly discontinuously and co-compactly by isometries does not contain any isometrically embedded flat planes, then G is hyperbolic.*

Question 2.1.8 (Gromov Question). Gromov asked if the local (CAT(0) or CAT(-1)) notion of hyperbolic group is in fact equivalent to the course notion (Gromov hyperbolic). We phrase this as two questions below. Clearly a positive answer to the first implies a positive answer to the second. Currently, both questions are open.

1. Is every Gromov hyperbolic group also CAT(-1)?

2. Is every Gromov hyperbolic group also CAT(0)?

The next theorem concerns the isoperimetric behavior of CAT(0) groups. The two basic examples of CAT(0) spaces, the hyperbolic plane and the Euclidean plane, have linear and quadratic isoperimetric functions respectively. This is the full range of isoperimetric behavior of CAT(0) groups.

Theorem 2.1.9 (Dehn functions on CAT(0) groups). *If G is a CAT(0) group, then the Dehn function of G is either linear or quadratic.*

In particular the snowflake groups of the previous chapter are not CAT(0) groups, because their Dehn functions are too large. The next theorem shows why the snowflake groups are not even subgroups of CAT(0) groups.

Theorem 2.1.10 ($\mathbb{Z}$ subgroups of CAT(0) are undistorted). *Every infinite cyclic subgroup of a* CAT(0) *group is undistorted.*

2.1.2 M_κ-complexes, the link condition

In practice one generally builds non-positively curved spaces out of convex cells in some model space. There is a local condition (the link condition) which ensures that such spaces are indeed locally CAT(0). We shall give informal definitions of these spaces and their links, and then give a careful statement of the link condition. The reader can consult [17] for detailed definitions.

Informally, a *piecewise Euclidean (resp. piecewise hyperbolic, piecewise spherical) cell complex K* is a complex obtained from a collection of convex cells in Euclidean space (resp. hyperbolic space, the sphere) by identifying their faces via isometries. The formal definition is similar to the charts definition for a geometric structure on a manifold, but we will not give it here.

In order to describe the local (link) structure of these M_κ-complexes, we need to define links of vertices of convex cells in the model spaces. If v is a vertex of a convex n-cell c in some model space, then the *link of v in c* (denoted by $Lk(v, c)$) is defined as the collection of unit tangent vectors to the space at v which point into c. This is a spherical cell of dimension $(n-1)$. Let c and c' be two convex cells which are identified along faces via an isometry f. Suppose that $v \in c$ and $v' \in c'$ are vertices which get identified by f. The derivative of f gives a spherical isometry which identifies a face of $Lk(v, c)$ with a face of $Lk(v', c')$. Thus, the *link of a vertex, v,* in an M_κ-complex, K, is naturally a *piecewise spherical complex*. We denote this link by $Lk(v, K)$.

The following local condition guarantees when an M_κ-complex is locally CAT(κ).

Theorem 2.1.11 (Link Condition). *The M_κ-complex K is locally* CAT(κ) *if the link of every vertex in K is* CAT(1).

This is a recursive condition to check, because the links of vertices are piecewise spherical complexes of lower dimension. In order to see that these links are CAT(1), one needs to verify two things:

1. Check that the links are locally CAT(1), via a repeated application of the link condition to these piecewise spherical complexes.

2. Check that the links do not contain any geodesic loops of length less than 2π.

In practice it is very hard to implement the link condition. The hard part occurs in checking part 2 above; namely, in showing that an arbitrary piecewise spherical complex does not contain any short geodesic loops. See [22] and [23] for interesting approaches to this question, and [24] for an application to 3-manifold groups.

There are two situations (and some variations on these), where verifying that there are no short geodesic loops in a piecewise spherical complex is easy to carry out.

1. First is when the spherical complex is 1-dimensional. In this case, one just checks that the metric graph has no essential loops of length less than 2π. Such graphs are called "large".

   To summarize, a 2-dimensional M_κ-complex, K, is locally CAT(κ) if $Lk(v, K)$ is large for all vertices $v \in K$.

2. Second is when the spherical complex is composed of *right angled spherical simplices*. An n-dimensional right angled spherical simplex is defined to be (isometric to) the span of the $n + 1$ standard unit basis vectors on the unit sphere in $\mathbb{R}^{n+1}$.

   In this case, there is a wonderful result due to Gromov (Gromov's Lemma) which reduces the CAT(1) check to a purely combinatorial one. *A piecewise spherical complex consisting of right-angled spherical simplices is* CAT(1) *if and only if it is a flag complex (clique complex).* The definition of a flag complex is given below.

   Piecewise Euclidean cubical complexes constitute an important family of examples of M_κ-complexes which have links composed of right-angled spherical simplices. The condition that a piecewise Euclidean cubical complex, K, be locally CAT(0) is that $Lk(v, K)$ is a flag complex for every vertex $v \in K$. We shall consider these complexes in more detail in the next section.

Definition 2.1.12 (Flag complex). A simplicial complex K is said to be a *flag complex* if every collection of pairwise adjacent vertices of K spans a simplex of K.

Ian Leary refers to this as the *every non-simplex contains a non-edge* criterion. Note that in the case the links are all 1-dimensional and right-angled, then the edge lengths are all exactly $\pi/2$. The flag condition ensures that there are no circuits of length 3 or less, and so the shortest embedded loops are of length at least 2π. Thus, the flag condition implies that the links are "large".

Example (Examples for discussion). For each of the finite presentations listed below, examine if the standard presentation 2-complex admits a locally CAT(0) structure. If it does not, can the 2-complex (staying in same homotopy class) be altered to get a locally CAT(0) complex? For the complexes listed below, do they admit CAT(0) structures, and if they do not, explain why not.

1. $\langle a, b | ab = ba \rangle$

2. $\langle a, b | aba = b \rangle$

3. $\langle a, b | abb = baa \rangle$

4. $\langle a, b | aba = bab \rangle$

5. $\langle a, b | abaa = bb \rangle$

6. $\langle a, b | aba = bb \rangle$

7. $\langle a, b | ab = bba \rangle$

8. $\langle a, b, x, y | [a, x] = [b, x] = [a, y] = [b, y] = 1 \rangle$

9. Does the 2-disk admit a CAT(0) metric?

10. Same question for the 2-sphere.

11. Same question for the duncehat.

2.1.3 Piecewise Euclidean cubical complexes

Recall that one major goal of this chapter is to construct examples of CAT(0) groups which contain subgroups with large Dehn functions. We shall construct three sets of examples; subgroups of free-by-cylic times free groups in Section 2.3, subgroups of right angled Artin groups in Section 2.5, and a subgroup of a hyperbolic group in Section 2.6. In all three cases, the ambient groups are CAT(0) cubical groups. In this subsection we give a more careful definition of piecewise Euclidean cubical complexes, and explore the non-positive curvature condition. Recent work of Wise, [37], demonstrates that the class of CAT(0) cubical groups is larger than might have been originally suspected, and now includes large families of small cancellation groups.

Definition 2.1.13 (Standard cubes and their faces). Given a non-negative integer m, a *standard m-cube* $\square^m$ is a copy of the product $[0, 1]^m$ in $\mathbb{R}^m$ with the usual Euclidean product metric. By composing this inclusion map with an arbitrary isometric embedding of $\mathbb{R}^m$ into $\mathbb{R}^n$ (for $n \geq m$) we can think of m-cubes as lying in high dimensional Euclidean spaces.

Let $0 \leq k \leq m$ be an integer. By a *k-dimensional face* of the standard m-cube $\square^m$ we mean a product $J^m \subset \square^m$ where $J = [0, 1]$ for k of the factors, and J is either $\{0\}$ or $\{1\}$ for each of the remaining $(m - k)$ factors.

Definition 2.1.14 (Cell complexes). Recall the definition of cell complexes (e.g., from Allen Hatcher's book *Algebraic Topology* [30]). The key idea is that these complexes X are build up inductively one dimension at a time. Start with a discrete collection of points, called the 0-skeleton, $X^{(0)}$. Now one builds up the k-skeleton, $X^{(k)}$, from the $(k-1)$-skeleton, $X^{(k-1)}$, as follows. Extra data: a disjoint collection of k-balls, and *attaching maps*. These are called the *k-cells* of X. Each ball comes with a continuous map (its attaching map) from its boundary $(k - 1)$-sphere to $X^{(k-1)}$. The k-skeleton is defined to be the quotient of the disjoint union of $X^{(k-1)}$

and the collection of k-balls by the equivalence relation defined by the attaching maps. The complex X is taken to be the (possibly infinite dimensional) union of the skeleta $X^{(k)}$, and is given the weak topology. For each k-cell of X, the "inclusion compose quotient" induced map $B^k \to X$ is called the *characteristic map* of the cell.

Definition 2.1.15 (Finite dimensional PE cubical complexes). Informally, a (finite dimensional) *piecewise Euclidean cubical complex* is a cell complex which is obtained from a collection of standard cubes of dimension at most n by identifying their faces via isometries. These are called PE cubical complexes for short. This gluing by isometries is captured in the formal definition below with the aid of admissible characteristic maps. This is reminiscent of the "local charts plus restricted family of transition maps" definition of a Riemannian manifold (or of a Euclidean or hyperbolic manifold etc).

   Formally, a (finite dimensional) *piecewise Euclidean cubical complex* X is a cell complex with the following additional structure.

1. [The cells] There exists a positive integer n such that each cell of X is a standard m-cube, for some $m \leqslant n$.

2. [Admissible maps] Given an m-cell $e \subset X$ with characteristic map

$$f_e : \square_e^m \to X$$

   and an isometry φ of $\mathbb{R}^n$, the composition

$$f_e \circ \varphi : \varphi^{-1}(\square_e^m) \to X$$

   is called an *admissible characteristic map* for the m-cell e.

3. [The gluing maps] For each m-cell e of X, the restriction of the characteristic map f_e to any k-face of $\square_e^m$ is an admissible characteristic map for a k-cell of X.

Definition 2.1.16 (Links in standard m-cubes). Let v be a vertex (that is a 0-dimensional face) of a standard m-cube $\square^m$. The *link of v in $\square^m$* is defined to be the set of inward pointing unit tangent vectors to $\square^m$ at v; that is, those unit tangent vectors at v which are contained in $\square^m$. It is a right angled spherical $(m-1)$-simplex, and we denote it by $Lk(v, \square^m)$.

Definition 2.1.17 (Links in PE cubical complexes). Let $x_0 \in X$ be a vertex in a PE cubical complex X. We say that the m-cell $e \subset X$ *contains x_0* if there is a vertex $v \in \square_e^m$ such that the characteristic map $f_e : \square_e^m \to X$ takes v to x_0.

   The *link of x_0 in X* is denoted by $Lk(x_0, X)$ and is defined to be a quotient space of the (disjoint) union of the right-angled spherical simplices $Lk(v, \square_e^m)$ where e is an m-cell of X ($m \geqslant 1$) which contains x_0, and $v \in \square_e^m$ is such that $f_e(v) = x_0$. Note that a given m-cell $e \subset X$ can contribute up to as many as 2^m distinct $(m-1)$-simplices to $Lk(x_0, X)$.

The quotient map is defined as follows. The $(r-1)$-simplex $Lk(v_1, \square^r_{e_1})$ is identified with a face of the $(s-1)$-simplex $Lk(v_2, \square^s_{e_2})$ if the characteristic map f_{e_2} restricts to an r-dimensional face of $\square^s_{e_2}$ (containing v_2) to give an admissible characteristic map of the cell e_1.

Remark 2.1.18. This seems like a mouthful, but it's pretty easy to get after you work a few specific examples. Again the intuition is easier to state than the formal definition. *Each m-cell of X contributes 2^m simplices of dimension $(m-1)$ to the links of various vertices of X, and these right-angled spherical simplices are glued together by isometries of their faces which are just obtained by differentiating the isometries used to glue the original collection of cubes together.*

Remark 2.1.19 (Hatcher's Δ-complexes). The links of vertices in PE cubical complexes are not quite simplicial complexes. They are examples of what Allen Hatcher calls Δ-complexes in his book, *Algebraic Topology* [30].

Example (The 2-torus). The quotient space of the unit square obtained by identifying opposite edges by translations gives a PE cubical 2-complex structure to the 2-torus T^2. This is also known as a PE squared structure. The quotient torus consists of one vertex (x_0 say), two 1-cells, and one 2-cell. It is easy to see that each 1-cell contributes 2-vertices to $Lk(x_0, T^2)$, and that the single 2-cell contributes exactly four 1-cells to $Lk(x_0, T^2)$. A little further thought enables us to conclude that $Lk(x_0, T^2)$ is isomorphic to a circle subdivided into 4 arcs. Metrically, each arc has a spherical structure of length $\pi/2$, so that the whole circle has length 2π.

Definition 2.1.20 (The flag condition for Δ-complexes). A Δ-complex (think of a typical link above, or look in Hatcher's book [30]) is said to be *flag* if

1. it is a simplicial complex, and

2. every collection of m pairwise adjacent vertices span an $(m-1)$-simplex.

Remark 2.1.21. The "simplicial" condition states that each pair of simplices intersects in at most a common face.

Definition 2.1.22 (Spherical complex construction). Let K be a finite simplicial complex. The *spherical complex associated to K*, denoted by $S(K)$, is defined by replacing every cell of K by an appropriately triangulated sphere of the same dimension. Here is the precise construction.

Let $\{v_1, \ldots, v_n\}$ be the vertices of K. The $S(K)$ has vertex set consisting of the set $\{v_1^+, v_1^-, \ldots, v_n^+, v_n^-\}$. Thus, each 0-simplex $\{v_i\}$ of K corresponds to a 0-sphere $\{v_i^+, v_i^-\}$ in $S(K)$. An m-simplex, τ, of K is just the join

$$\tau = \{v_{i_0}\} * \cdots * \{v_{i_m}\}$$

of a collection of $(m+1)$ vertices of K. The corresponding m-sphere, denoted by $S(\tau)$, in $S(K)$ is defined to be the join

$$S(\tau) = \{v_{i_0}^-, v_{i_0}^+\} * \cdots * \{v_{i_m}^-, v_{i_m}^+\}$$

Example (Exercises on the Spherical complex construction). Describe/draw the complex $S(K)$ in the following situations.

1. K is a line segment of length 1 (one 1-simplex).

2. K is a line segment of length 2 (two 1-simplices).

3. K is a pair of 2-simplices meeting at a vertex.

4. K is a pair of 2-simplices meeting along an edge.

5. K is the boundary of a square. (Draw $S(K)$.)

Now, prove that if K is a simplicial complex, then $S(K)$ is also a simplicial complex.

Example (Exercises on flag condition). These exercises require the reader to recall (or look up) the definitions of *join, barycentric subdivision* and *link of a simplex*.

1. Prove that the join of two flag complexes is again flag.

2. Prove that the barycentric subdivision of any simplicial complex is flag.

3. Prove that the link (in sense of simplicial complexes) of any simplex in a flag complex is again a flag complex.

4. Prove that if K is flag, then the associated spherical complex $S(K)$ is also flag.

Definition 2.1.23 (Non-positive curvature — Gromov's condition). A PE cubical complex X is said to be *non-positively curved* if it satisfies the "Gromov flag condition"; namely, $Lk(x_0, X)$ is flag for each $x_0 \in X^{(0)}$.

Example (Exercise). Prove that the direct product of two non-positively curved PE cubical complexes is also a non-positively curved PE cubical complex.

2.2   Morse theory I: recognizing free-by-cyclic groups

In the first subsection we introduce the notions of Morse functions on affine cell complexes. In the second subsection we give a Morse theory criterion which guarantees that a 2-complex is aspherical, and has free-by-cyclic fundamental group. This criterion has been used by Howie [31] in the context of 2-complexes associated to LOT groups. It has also been used in [1] to build examples of CAT(0) free-by-cyclic groups which contain closed hyperbolic surface subgroups with arbitrary polynomial or exponential distortion.

2.2.1   Morse functions and ascending/descending links

Intuitively, a Morse function on a cell complex is a real-valued function on the cell complex with the property that when we restrict it to a cell, it obtains a unique maximum and a unique minimum, and these are attained at vertices of the cell. One should think of holding a square (or a cube) by one vertex and letting it hang freely. The usual height function gives a good intuition of a Morse function restricted to these cells.

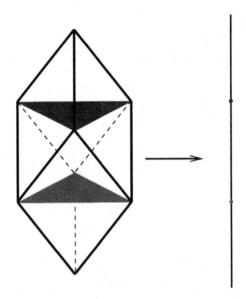

Figure 2.1: Typical Morse function on a 3-cube.

In these lectures we shall assume that the Morse functions have the property that the image of the 0-skeleton of the domain complex is a discrete subset of $\mathbb{R}$. The definitions and theory are taken from [4]. There is a nice treatment of more general Morse functions in [20]. See [12] for applications of tree-valued Morse functions on simplicial complexes.

Consider the preimage of a real number t under a Morse function f. This is called a t-level set. If t is not in the image of the 0-skeleton, then by discreteness, there is a small neighborhood $(t - \epsilon, t + \epsilon)$ of t, such that $f^{-1}((t - \epsilon, t + \epsilon))$ is homeomorphic to $f^{-1}(\{t\}) \times (-\epsilon, \epsilon)$. Thus the topology of level sets does not change locally as long as we avoid the 0-cells. At 0-cells there can be non-trivial changes in topology.

For example, take as our domain cell complex the 2-skeleton of the 3-cube together with 2 squares as shown in Figure 2.2. The level set containing points just below the vertex v is a small circle consisting of a circuit of three 1-cells together with two edges attached to two of the vertices of the circle. Note that these level sets are homeomorphic for small perturbations of the target point. The level set containing the vertex v is just a segment of length 2. The former level sets have non-trivial π_1, while the latter level set is contractible, so there has been a definite

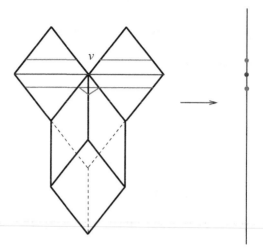

Figure 2.2: Morse function on a 2-complex with a circle as descending link.

change in topology. One way of expressing this is to say that the midpoint of the level set through v is homotopy equivalent to a cone on the circle, or that the circle in the lower level sets has been "coned off" by the vertex v. This change in topology (the coned off circle) is contained in the collection of 2-cells (squares) on which the restriction of the Morse function has a maximum at v. This local coning process will be encoded in the notions of *ascending and descending links* and will be formalized in the Morse Lemma from [4].

First we shall define a good collection of complexes on which one can define Morse functions, and then give the definitions of ascending and descending links.

Definition 2.2.1 (Affine cell complexes). A finite-dimensional cell-complex X is said to be an *affine cell-complex* if it is equipped with the following structure. An integer $m \geqslant \dim(X)$ is given, and for each cell e of X we are given a convex polyhedral cell $C_e \subset \mathbb{R}^m$ and a characteristic function $\chi_e : C_e \to e$ such that the restriction of χ_e to any face of C_e is a characteristic function of another cell, possibly precomposed by a partial affine homeomorphism (that is, the restriction of an affine homeomorphism) of $\mathbb{R}^m$.

Definition 2.2.2 (Morse function). A map $f : X \to \mathbb{R}$ defined on an affine cell-complex X is a *Morse function* if

- for every cell e of X, $f\chi_e : C_e \to \mathbb{R}$ extends to an affine map $\mathbb{R}^m \to \mathbb{R}$, and $f\chi_e$ is constant only when $\dim e = 0$, and

- the image of the 0-skeleton is discrete in $\mathbb{R}$.

Definition 2.2.3 (Circle-valued Morse function). A *circle-valued Morse function* on an affine cell complex X is a cellular map $f : X \to S^1$, with the property that f lifts to a Morse function between universal covers.

Definition 2.2.4 (Ascending and descending links). Suppose X is an affine cell complex and $f : X \to S^1$ is a circle-valued Morse function. Choose an orientation of S^1, which lifts to one of $\mathbb{R}$, and lift f to a map of universal covers $\tilde{f} : \tilde{X} \to \mathbb{R}$. Let $v \in X^{(0)}$, and note that the link of v in X is naturally isomorphic to the link of any lift $\tilde{v}$ of v in $\tilde{X}$. We say that a cell $\tilde{e} \subset \tilde{X}$ contributes to the *ascending* (respectively *descending*) link of $\tilde{v}$ if $\tilde{v} \in \tilde{e}$ and if $\tilde{f}|_{\tilde{e}}$ achieves its minimum (respectively maximum) value at $\tilde{v}$. The ascending (respectively descending) link of v is then defined to be the subset of $Lk(v, X)$ naturally identified with the ascending (respectively descending) link of $\tilde{v}$.

Example (Figure 2.2 revisited). In the example of Figure 2.2 the ascending link of the vertex v consists of a 0-sphere. This 0-sphere corresponds to the two 1-cells of the ambient complex which have a common minimum height at v. The descending link consists of the circle (union of three 1-cells). This circle corresponds to the union of the three closed 2-cells (squares) which have a common maximum height at v. There is a close connection between the topology of these ascending and descending links, and the topology of the three level sets; one through v, one slightly above v, and one slightly below v.

The level set through v is contractible. It is the union of two line segments attached at the common endpoint v. We denote the height (Morse) function by f, and choose coordinates on the real line so that the level set through v is the 0-level set, $f^{-1}(0)$.

Let $\epsilon > 0$ be small. The level set slightly above v, or the ϵ-level set, is the union of two disjoint line segments. This corresponds to the fact that the ascending link at v is disconnected (it is a 0-sphere). There is a "copy" of this 0-sphere ascending link in the ϵ-level set; it is the intersection of the ϵ-level set with the union of the two 1-cells of the ambient complex which have a common minimum height at v. The block $f^{-1}([0, \epsilon])$ contains a copy of the ϵ-level set together with this 0-sphere coned off to the vertex v. This complex (ϵ-level set with coned off 0-sphere) is homotopy equivalent to $f^{-1}([0, \epsilon])$. An explicit homotopy equivalence can be realized by "pushing up" through the ambient 2-complex.

Similarly, the $(-\epsilon)$-level set slightly below the vertex v is homotopy equivalent to a circle. This corresponds to the fact that the descending link at v is a circle. There is a "copy" of this circle descending link in the $(-\epsilon)$-level set; namely, the intersection of the $(-\epsilon)$-level set and the union of the three closed squares (2-cells) in the ambient complex which all attain a maximum height at v. The block $f^{-1}([-\epsilon, 0])$ contains a copy of the $(-\epsilon)$-level set together with this copy of the circle coned off to v. This complex (the $(-\epsilon)$-level set with coned off circle) is homotopy equivalent to $f^{-1}([-\epsilon, 0])$. An explicit homotopy equivalence is obtained by "pushing down" through the ambient 2-complex.

The next theorem formalizes these connections between the homotopy types of level sets through a vertex v on one hand, and the homotopy types of level sets just above (resp. below) the vertex v and the homotopy type of ascending (resp. descending) links on the other hand. This theorem appears as Lemma 2.5 of [4]. It connects the local ascending/descending links picture with the more global viewpoint of how the topology of the preimage $f^{-1}([a, b])$ changes as the interval $[a, b]$ increases so that the preimage includes more and more vertices.

Theorem 2.2.5 (Morse Lemma). *Let $f : X \to \mathbb{R}$ be a Morse function on an affine cell complex as above. Suppose $J \subset J' \subset \mathbb{R}$ are closed intervals, $\inf J = \inf J'$, and $J' \setminus J$ contains only one point r of f(0-cells). Then $f^{-1}(J')$ is homotopy equivalent to $f^{-1}(J)$ with the copies of $Lk_\downarrow(v, X)$ (v a vertex with $f(v) = r$) coned off.*
A similar statement holds when $\inf J = \inf J'$ is replaced by $\sup J = \sup J'$ and $Lk_\downarrow(v, X)$ by $L_\uparrow(v)$.

Remark 2.2.6. We refer the reader to [4] for the proof of this theorem. It is a good exercise to explicitly construct these homotopy equivalences onto the coned off complexes in the particular example of Figure 2.2.

2.2.2   Morse function criterion for free-by-cyclic groups

The next proposition gives a local way of telling that a 2-complex is aspherical and has free-by-cyclic fundamental group.

Proposition 2.2.7 (Free-by-cyclic). *If $f : X \to S^1$ is a circle-valued Morse function on the 2-complex X all of whose ascending and descending links are trees, then X is aspherical, and $\pi_1(X)$ is free-by-cyclic.*

Remark 2.2.8. This is basically a result of J. Howie [31]. In [31] it is proved that if all the ascending links (or all the descending links) of a given LOT 2-complex are trees, then the 2-complex is aspherical. Also, if both ascending and descending links are trees, then the LOT 2-complex has free-by-cyclic fundamental group.

The current proposition says that if all of both the ascending and the descending links of a general affine 2-complex are trees, then the 2-complex is aspherical and has free-by-cyclic fundamental group. The current formulation is given in [1].

Note that if just the ascending links are trees and the descending links are not, then the aspherical 2-complex need not have free-by-cyclic fundamental group. As a simple example, take the group with five generators t and a_i ($i = 0, \ldots, 3$) and four conjugation relations $a_i a_{i+1} a_i^{-1} = t$ ($i \mod 4$). In any map to $\mathbb{Z}$, all five generators must have the same image, so we can assume this common image is a generator of $\mathbb{Z}$. The corresponding circle-valued Morse function takes the a_i and the t loops once around a target circle. Its ascending link is a tree, but the descending link is the disjoint union of a single vertex and a circle. The kernel of the map to $\mathbb{Z}$ is not finitely generated.

Proof of Proposition 2.2.7. Look at generic f point preimages in X. Since X has an affine structure and $f : X \to S^1$ is affine, the point preimages are affine graphs in X. We may then view X as the total space of a graph of spaces: the underlying space is the target circle, the edge space is the generic point preimage graph, and the vertex space is the S^1-vertex preimage. The maps from an edge space to the adjacent vertex spaces collapse either the ascending or the descending subgraphs of the corresponding link. These are trees by hypothesis, so these maps are homotopy equivalences. It follows that X is homotopy equivalent to a graph bundle over S^1, and therefore is aspherical and has free-by-cyclic fundamental group. □

Example (Geisking manifold group). Consider the group with presentation

$$\langle a, b \,|\, abb = baa \rangle.$$

The link of the single vertex in the presentation 2-complex is the complete graph on the set $\{a^+, a^-, b^+, b^-\}$.

We shall apply the Morse criterion to see that it is free by cyclic. Suppose that the map to $\mathbb{Z}$ takes a to A and b to B. The relation becomes $A + 2B = B + 2A$. In other words, $A = B$, which we take to be a generator of $\mathbb{Z}$. So the picture of a typical lift of the relator 2-cell to the universal covering from the perspective of the Morse function is a hexagon with one vertex as its maximum point, and the diametrically opposite vertex as its minimum, and the two positive words abb and baa reading from the minimum point to the maximum point along either side of the 2-cell. Thus, the ascending link is the segment from a^- to b^-, and the descending link is the segment from a^+ to b^+. Since these are both trees, we conclude that the group is free-by-cyclic.

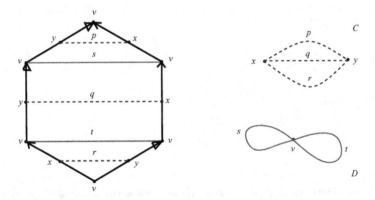

Figure 2.3: The 2-cell of the Geisking group, and the two preimage graphs.

Following the proof of Proposition 2.2.7, we see that the preimage of the base point in the target circle is a bouquet of two circles. This is the graph D of Figure 2.3. Hence the group is $F_2 \rtimes \mathbb{Z}$. One can also check that the preimage of a generic point on the circle, is a graph with 2 vertices and 3 edges. This is shown as the graph C in Figure 2.3. This graph is homotopy equivalent to the

bouquet of two circles, and two explicit homotopy equivalences are obtained by
letting the generic point on the circle tend toward the base point (there are two
directions to approach the base point). These homotopy equivalences can be easily
read off the 2-cell on the left of Figure 2.3. One homotopy equivalence collapses
the edge r to the point v, maps the edge p to the loop s, and maps the edge q
to the loop t (and is seen by sliding the dashed edges in the 2-cell down). The
second homotopy equivalence collapses the edge p to the point v, maps the edge q
to the loop s, and maps the edge r to the loop t (and is seen by sliding the dashed
edges in the 2-cell up). You should orient the edges to get a precise description of
the homotopy equivalences, and then you can read off the automorphism in the
$F_2 \rtimes \mathbb{Z}$ structure by composing one homotopy equivalence with an inverse of the
other. This describes the presentation 2-complex as the total space of a graph of
spaces, whose underlying graph is a circle with one vertex. The vertex space is the
bouquet D, and the edge space is the generic preimage graph C. The maps from
the edge to the vertex space are the two homotopy equivalences above.

Remark 2.2.9 (Level sets in universal cover perspective). There is another way to
look at this example. We know that the 2-complex X is non-positively curved by
the link condition. Thus its universal cover is CAT(0), and so contractible. The
circle-valued Morse function lifts to a Morse function $\tilde{f}$ on the universal cover of
X.

We know by the Morse Lemma, Theorem 2.2.5, that this contractible 2-
complex is built up from a level set $\tilde{f}^{-1}(t)$ by homotopy equivalences and repeatedly coning off copies of the ascending and descending links. But these are all trees,
so the coning operations are also homotopy equivalences. Thus the level set $\tilde{f}^{-1}(t)$
is homotopy equivalent to the CAT(0) universal cover of X, and so is contractible
too. But the level set is a graph, and therefore must be a tree.

Now, the kernel of the map $G \to \mathbb{Z}$ acts freely and cocompactly on $\tilde{f}^{-1}(t)$
(by restriction of its deck transformation action on all of the universal cover of
X). We conclude that the kernel is a finite rank free group.

This is a very fundamental viewpoint to which we shall return repeatedly in
later sections. We summarize the ideas here.

1. We are given a real-valued Morse function f on a contractible cell complex X.
   The contractibility of X is often a consequence of some type of non-positive
   curvature, such as the fact that X might be CAT(0) or CAT(-1).

2. Suppose that there is a group G acting cocompactly by deck transformations
   on the complex X. Suppose that there is an epimorphism $\varphi : G \to \mathbb{Z}$. Let $\mathbb{Z}$
   act cocompactly by translations on $\mathbb{R}$, and suppose that the map $f : X \to \mathbb{R}$
   is φ-equivariant with respect to the action of G on X and of $\mathbb{Z}$ on $\mathbb{R}$.

3. Local ascending/descending link information together with contractibility of
   X enable one to deduce topological properties of level sets $f^{-1}(t)$.

4. Now $\mathrm{Ker}(\varphi)$ acts cocompactly by restriction of deck transformations on a
   level set $f^{-1}(t)$. The topology of the level set will have consequences for the

kernel. In the previous example, the level sets are trees, and so the kernel is a free group. In examples of later sections, we shall encounter situations where the kernel may be finitely generated but not finitely presentable, or finitely presented but not of type F_3.

Example (Morse Criterion Examples). Run the Morse criterion on the following group presentations. Say what the rank of the free kernel is. Write out the automorphism of the free group explicitly.

1. $\langle a, b \mid aba = bab \rangle$

2. $\langle a, b \mid abaa = bb \rangle$

3. $\langle a_0, \ldots, a_n \mid a_i^{a_{i+1}} = a_0 \ (1 \leqslant i \leqslant n - 1), \ a_0^{a_1} = a_n \rangle$.

2.3 Groups of type $(F_n \rtimes \mathbb{Z}) \times F_2$

In this section we describe a family of 3-dimensional CAT(0) cubical groups which have subgroups with polynomial Dehn functions of arbitrary degree. The examples are obtained using the Bieri doubling trick, [5] and [2], on a family of CAT(0) squared free-by-cyclic groups. These groups are described as LOG groups. We define LOT/LOG groups in the first subsection below. We describe the free-by-cyclic examples in the second subsection. In the third subsection, the Bieri trick is introduced and the 3-dimensional CAT(0) cubical examples are given. These examples are very close to a family of CAT(0) groups and polynomial Dehn function subgroups given by Martin Bridson [16]. The only difference is that the examples below are CAT(0) cubical, whereas Martin's examples are CAT(0) with 3-cells which are products of 2-simplices and 1-simplices.

2.3.1 LOG groups and LOT groups

Definition 2.3.1 (LOG groups and LOT groups). A *labeled, oriented graph*, or LOG, consists of a finite, directed graph with distinct labels on all the vertices, and oriented edge labels taken from the set of vertex labels. An LOG defines a finite presentation as follows. The set of generators is in 1–1 correspondence with the set of vertex labels, and the set of relators is in 1–1 correspondence with the set of edges, so that an edge labeled a oriented from vertex u to vertex v corresponds to a conjugation relation $ava^{-1} = u$. In case the graph is a tree we call it a *labeled, oriented tree* or LOT and call the corresponding finite presentation an LOT presentation.

Note that the presentation 2-complex of an LOG group is a squared complex. In the examples that follow, the squared complexes are non-positively curved.

Remark 2.3.2 (Why consider LOTs). People who study low dimensional topology are interested in LOTs, mainly because they describe 2-complexes K_T which are useful in working on *Whitehead's Conjecture*. This conjecture states that *a sub-complex of a contractible 2-complex is an aspherical 2-complex* (i.e. its universal cover is contractible).

Where do LOTs come into all this? Well, if you take a LOT complex K_T and attach a disk to any of its 1-cell circles, the result is always contractible (seeing this is a nice exercise in algebraic topology). So, if an LOT complex K_T were not aspherical, then it would provide a counterexample to the Whitehead Conjecture. According to Jim Howie, the LOTs form the most important special case of the (finite) Whitehead conjecture. In fact, modulo another conjecture in low dimensional topology (Andrews-Curtis Conjecture), Howie has proved that the finite Whitehead conjecture reduces to testing that all LOT complexes are aspherical.

2.3.2   Polynomially distorted subgroups

In this subsection we describe CAT(0) groups of the form $F_n \rtimes \mathbb{Z}$ where the F_n subgroup has polynomial distortion of degree n. Recall that the *distortion function* f of a finitely generated subgroup H of a finitely generated group G measures roughly how much the n-ball in G gets distorted in H. It is defined precisely as follows

$$f(n) \;=\; \max\{\, d_H(1,h) \,|\, h \in H, d_G(1,h) \leqslant n \,\}$$

where d_H and d_G are the metrics on H and G respectively. The distortion function f depends up to equivalence only on the pair (G, H), and not on the choice of finite generating sets.

Let $\varphi : F_n \to F_n$ be the automorphism that is defined on the free basis $\{a_1, \ldots, a_n\}$ as follows:

$$\varphi(a_j) = a_j a_{j-1} \quad (j \text{ odd}), \qquad \varphi(a_j) = a_{j-1} a_j \quad (j \text{ even})$$

where a_0 denotes the identity element 1.

Now consider the group $F_n \rtimes_\varphi \mathbb{Z}$ where $\mathbb{Z} = \langle \tau \rangle$. Define α_i by

$$\alpha_i a_i \;=\; \tau \quad (\text{for } i \text{ odd}), \quad \text{and} \quad \alpha_i \;=\; a_i \tau \quad (\text{for } i \text{ even}).$$

Then $F_n \rtimes_\varphi \mathbb{Z}$ is isomorphic to the group with the LOG presentation shown in Figure 2.4. The graph has $n + 1$ vertices labeled τ and α_i for $1 \leqslant i \leqslant n$. The vertex α_n is isolated, there is a loop labeled α_1 at τ, and directed edges α_i from α_{i-1} to τ for $2 \leqslant i \leqslant n$. Figure 2.4 also shows the link of the single vertex in the presentation 2-complex. This link is large, and so the 2-complex is non-positively curved. We define a circle-valued Morse function on this complex by sending each generator around the circle once in the positive direction and extending linearly over the 2-cells. This labels the vertices with $+$ or $-$ as shown in Figure 2.4.

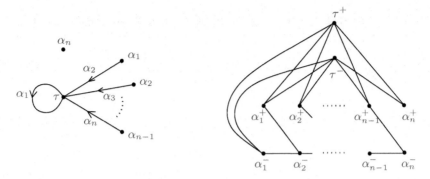

Figure 2.4: LOG presentation of $F_n \rtimes_\varphi \mathbb{Z}$, and link of vertex in the associated squared 2-complex.

Proposition 2.3.3 (Polynomially distorted kernels). *Let $\varphi : F_n \to F_n$ be the automorphism defined on a free basis $\{a_1, \ldots, a_n\}$ of F_n as*

$$\varphi(a_j) = a_j a_{j-1} \quad (j \text{ odd}), \qquad \varphi(a_j) = a_{j-1} a_j \quad (j \text{ even})$$

where $a_0 = 1$. Then the free fiber F_n has polynomial distortion of degree n in $F_n \rtimes_\varphi \mathbb{Z}$.

Sketch of proof. We need to argue upper and lower bounds for the distortion of F_n in $F_n \rtimes_\varphi \mathbb{Z}$. Let t denote the stable letter in the HNN description of $F_n \rtimes_\varphi \mathbb{Z}$.

1. *Upper Bounds.* First argue that $t^n a_j t^{-n}$ is a word whose length is of order n^{j-1} in the a_i. Likewise for $t^{-n} a_j t^n$. Now argue that any word w in the a_i and t which represents an element of F_n can be reduced to a word w' which does not involve t. Use your previous length estimates to argue that the length of w' is at most of order $|w|^n$. If w is a geodesic word, then we see that the distortion of F_n is at most polynomial of degree n.

2. *Lower Bounds.* Find explicit words in a_i and t which represent elements of F_n, and whose length (as a reduced word in the a_i) grows to $|w|^n$ on eliminating t's. Try $t^k a_n^k t^{-k}$. It is a word of length $3k$, which should "reduce" to a word in F_n of length at least k^n. □

Example (Area estimates). Take the word w that you used in the lower bound estimates above. Let w' denote the reduced word in F_n which represents the same element as w. What is the area of the word $w^{-1} w'$ (which represents the identity) in $F_n \rtimes_\varphi \mathbb{Z}$?

2.3.3 Examples: The double construction and the polynomial Dehn function

The *double of $F_n \rtimes \mathbb{Z}$ over F_n* is defined to be the free product with amalgamation of two copies of $F_n \rtimes \mathbb{Z}$ amalgamated along the common F_n.

Lemma 2.3.4 (Dehn function of the double). *Let $\varphi\colon F_n \to F_n$ be the automorphism of Proposition 2.3.3. The Dehn function of the double of $F_n \rtimes_\varphi \mathbb{Z}$ over the F_n kernel is $\delta(x) \sim x^{n+1}$.*

It is an exercise to give a careful proof of this, taking care of both upper and lower bounds. Here are some hints to help the reader. The double has the following presentation:

$$\langle\, a_1, \ldots, a_n, s, t \mid a_i^t = \varphi(a_i) = a_i^s \text{ for } 1 \leqslant i \leqslant n \,\rangle$$

Note that this is the fundamental group of a graph of groups with underlying graph a bouquet of two circles. The vertex and edge groups are all F_n and the edge to vertex maps are the identity and φ. This group has the free group on $\{s, t\}$ as a retract. We see this by noting that $F_{\{s,t\}}$ is also the fundamental group of a graph of groups with underlying graph a bouquet of two circles, only this time with trivial edge and vertex groups. The trivial group is obviously a retract of the free group F_n and this extends to the two graphs of groups to yield that $F_{\{s,t\}}$ is a retract of the double group. In particular, the subgroup of the double generated by $\{s, t\}$ is free on s and t, and is undistorted in the double group (verify this!).

1. *Upper Bounds.* The idea is to prove upper bounds by induction on n. To start the induction, note that the double

$$\langle\, a_1, s, t \mid a_1^s = a_1 = a_1^t \,\rangle$$

is just $F_2 \times \mathbb{Z}$ and so has quadratic Dehn function. In particular, its Dehn function is bounded above by x^{1+1}, and the base case is established.

To establish the inductive step, we cut van Kampen diagrams over $a_1, \ldots, a_n$, s, t along a_n-corridors. Note that the words along one side of an a_n-corridor are words in the group generated by $\{s, t\}$. By the remarks in the previous paragraph, this group is an undistorted copy of $F_{\{s,t\}}$ in the double group $F_{n-1} \rtimes F_{\{s,t\}}$. Therefore, there are no a_n-annuli in reduced van Kampen diagrams over $\{a_1, \ldots, a_n, s, t\}$.

In particular, each a_n-corridor begins and ends on the boundary of the van Kampen diagram, and so there are at most $|\partial|/2$ of these corridors (where $|\partial|$ denotes the boundary length of the diagram). Conclude that the original diagram over $a_1, \ldots, a_n$, s, t is cut into at most $|\partial|/2$ strips (a_n-corridors) and complimentary regions. Argue that each complimentary region is a van Kampen diagram over $a_1, \ldots, a_{n-1}$, s, t of boundary length at most $2|\partial|$. By induction this has area at most some constant times $(2|\partial|)^n$. Conclude that the original van Kampen diagram has area at most a constant times $|\partial|^{n+1}$.

2. *Lower Bounds.* From the exercise at the end of the previous subsection, one should deduce that the word

$$t^k a_n^k t^{-k} s^k a_n^{-k} s^k$$

has area at least k^{n+1}. Since its length is a linear function of k, the lower degree $(n+1)$ polynomial lower bound is established.

Theorem 2.3.5 (Subgroups of CAT(0)). *There exist 3-dimensional cubical CAT(0) groups which contain subgroups with Dehn function of the form $\delta(x) \sim x^m$ for each $m \geqslant 3$.*

Sketch of Proof. Just consider the groups $F_n \rtimes_\varphi \mathbb{Z}$ defined above. These are CAT(0) squared groups. Thus the groups $G_n = (F_n \rtimes_\varphi \mathbb{Z}) \times F_2$ are CAT(0) cubical groups. To see this just take the product of the non-positively curved squared complexes for $F_n \rtimes_\varphi \mathbb{Z}$ with a bouquet of two circles.

Let the generators of the F_2 factor be u and v. Check that the subgroup of G_n generated by $\{a_1, \dots, a_n, ut, vt\}$ is isomorphic to the double of $F_n \rtimes_\varphi \mathbb{Z}$ over the fiber F_n. Set $m = n + 1$. □

Remark 2.3.6 (Bieri–Stallings origin of doubling). The idea of embedding doubles in direct products is explored extensively in [2] and has its origin in the work of Bieri [5]. In turn [5] was inspired by the work of [36]. The doubling trick was Bieri's way of seeing how Stallings' example fit into a direct product of three free groups, and so led to the consideration of finiteness properties of subgroups of direct products of many free groups. We shall see a Morse theory way of understanding some of these subgroups in Section 2.4 below.

2.4 Morse theory II: topology of kernel subgroups

In this section we discuss a model situation which has many interesting applications and specifications. The model consists of a locally CAT(0) complex K and a circle valued Morse function $l \colon K \to S^1$ which induces an epimorphism $l_* \colon \pi_1(K) \to \mathbb{Z}$. Lifting l to the universal covers gives an equivariant Morse function $f \colon \widetilde{K} \to \mathbb{R}$. The group $\pi_1(K)$ acts by deck transformations on $\widetilde{K}$. The kernel subgroup $\mathrm{Ker}(\pi_1(K) \to \mathbb{Z})$ acts by restriction of deck transformations on a level set $f^{-1}(t)$ with quotient $l^{-1}(\bar{t}) \subset K$ where $t \in \mathbb{R}$ is in the preimage of $\bar{t} \in S^1$.

We investigate the relationship between local topology on one hand (the topology of ascending and descending links of the Morse function f), and global topology on the other hand (the equivariant topology of level sets $f^{-1}(t)$). We also explore relationships between these local and global topological data and finiteness properties of the kernel subgroups $ker(\pi_1(K) \to \mathbb{Z})$.

Detailed proofs of typical theorems which examine these relationships can be found in [4] and in [8]. The version we work with in these notes is given as Theorem 2.4.2 below. Instead of giving a general proof, we will focus on three instructive examples involving kernels of maps from products of free groups to $\mathbb{Z}$ which illustrate the techniques of the proof. The examples increase in dimension and in complexity of their kernel subgroups. However, there is a formal outline which the arguments in all three examples share.

On one hand, many similarities between the examples might be noted; the idea of working with a topological model, the existence of a Morse function on a highly connected (contractible in these 3 examples) complex, the reduction of finiteness properties of the kernel to topological properties of level sets, the Morse theory argument to prove the topological properties of the level sets.

On the other hand, one might remark at the increase in topological and combinatorial complexity of the level sets in the higher dimensional examples. Keep in mind that these are just "slices" of a product of trees. What is exciting, and yet to be fully understood, is the geometric structure of these level sets. They are "slices" of CAT(0) cubical complexes, but come with their own piecewise Euclidean metrics, and interesting geometries. What are their Dehn functions? Higher Dehn functions? They seem to have some type of self-similarity (this is the case for the kernels of right-angled Artin groups in the next section). What does this geometric self-similarity say about the algebra? What does it say about the large scale geometry of these kernels? For example, what can one say about their asymptotic cones? This is a rich area to explore.

This section is organized as follows. We begin with a definition of finiteness properties of groups, then give the statement of the local to global topology theorem. We work carefully through three explicit examples, and then finish with a tie in to the finiteness properties of the kernel group in Section 2.6.

Definition 2.4.1 (Finiteness property F_n). A group G is said to be *of type F_n* if it has an Eilenberg–Mac Lane complex $K(G, 1)$ with finite n-skeleton. Equivalently, a group is of type F_n if it acts freely, faithfully, properly, cellularly, and cocompactly on an $(n-1)$-connected cell complex.

Example (F_1 and F_2). Verify that a group is finitely generated if and only if it is of type F_1, and is finitely presented if and only if it is of type F_2.

Example (Finiteness properties of hyperbolic groups). A torsion free hyperbolic group has a finite Eilenberg–Mac Lane space. In particular, it is of type F_n for all integers $n \geqslant 1$. The quotient of the Rips complex by the torsion free hyperbolic group gives a finite Eilenberg–Mac Lane space.

Morse theory setup for Theorem 2.4.2.

1. We are given a real-valued Morse function f on a contractible cell complex X. The contractibility of X is often a consequence of some type of non-positive curvature, such as the fact that X might be CAT(0) or CAT(-1).

2. Suppose that there is a group G acting cocompactly by deck transformations on the complex X. Suppose that there is an epimorphism $\varphi : G \to \mathbb{Z}$. Let $\mathbb{Z}$

act cocompactly by translations on $\mathbb{R}$, and suppose that the map $f : X \to \mathbb{R}$ is φ-equivariant with respect to the action of G on X and of $\mathbb{Z}$ on $\mathbb{R}$.

Theorem 2.4.2. *Given the Morse theory setup above.*

1. *Suppose that $Lk_\downarrow(v, X)$ and $Lk_\uparrow(v, X)$ are $(n-1)$-connected for every $v \in X^{(0)}$, then $f^{-1}(t)$ is $(n-1)$-connected. Therefore, $\mathrm{Ker}(\varphi)$ is of type F_n.*

2. *Suppose X is $(n+1)$-dimensional, and $Lk_\uparrow(v, X)$ are $(n-1)$-connected for every $v \in X^{(0)}$. If there are vertices v with $H_n(Lk_\uparrow(v, X); \mathbb{Z}) \neq 0$ or $H_n(Lk_\downarrow(v, X); \mathbb{Z}) \neq 0$, then $\mathrm{Ker}(\varphi)$ is of type F_n but not of type F_{n+1}.*

We look at three special examples of this theorem in the next three subsections.

2.4.1   A non-finitely generated example: $\mathrm{Ker}(F_2 \to \mathbb{Z})$

Let $\varphi : F_{\{x,y\}} \to \mathbb{Z} : x, y \mapsto a$ where a is a generator of the infinite cyclic group $\mathbb{Z}$. We give a Morse theory argument to prove that $\mathrm{Ker}(\varphi)$ is not finitely generated.

Let $S^1 = \mathbb{R}/\mathbb{Z}$ have one 0-cell and one 1-cell. Start with a topological representative $l : S^1 \vee S^1 \to S^1$ where the wedge product has one 0-cell and two 1-cells. The map l maps the single 0-cell of $S^1 \vee S^1$ to the 0-cell of S^1, and is a homeomorphism when restricted to the open 1-cells of $S^1 \vee S^1$. The induced map, l_*, on π_1 is just the homomorphism φ. The lift of l to the universal cover gives a Morse function $f : T \to \mathbb{R}$ where T denotes the infinite 4-valent tree. This is pictured in Figure 2.5.

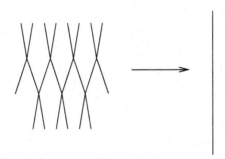

Figure 2.5: The Morse function on the tree T.

The ascending link of each vertex consists of a 0-sphere. This corresponds to the fact that there are two edges adjacent to a given vertex which are above the vertex. The descending link is also a 0-sphere. This corresponds to the fact that there are two edges adjacent to, but below, each vertex.

Pick an integer $t \in \mathbb{R}$. The level set $f^{-1}(t)$ is a collection of vertices of T which are all at height t. The kernel $\mathrm{Ker}(F_2 \to \mathbb{Z})$ acts on the level set $f^{-1}(t)$

with quotient equal to the one 0-cell of $S^1 \vee S^1$. To see this note that all of F_2 acts on T via deck transformations with quotient equal to $S^1 \vee S^1$. The kernel consists of precisely those deck transformations in F_2 which map to 0 in $\mathbb{Z}$; that is those deck transformations which do not translate up or down in the induced action on $\mathbb{R}$. How are we to argue from this situation that $\mathrm{Ker}(F_2 \to \mathbb{Z})$ is not finitely generated? For example, what is different about this situation than, say, the action of $\mathbb{Z}^2$ on the set of integer lattice points in the plane? In both situations we have a group acting transitively and freely on a discrete set of points.

In the case of $\mathbb{Z}^2$ it is possible to attach two $\mathbb{Z}^2$-equivariant families of 1-cells to the integer lattice in the plane, to get a connected 1-complex. For example one family consists of translates of a line segment from $(0,0)$ to $(1,0)$ (corresponding to one generator of $\mathbb{Z}^2$) and the other family consists of translates of a line segment from $(0,0)$ to $(0,1)$ (corresponding to a second generator of $\mathbb{Z}^2$).

Reduction to topological properties of the level set.

We have the following level set formulation of the non-finite generation of $H = \mathrm{Ker}(F_2 \to \mathbb{Z})$.

1. If $H_0(f^{-1}(t); \mathbb{Z})$ is not finitely generated as a $\mathbb{Z}H$-module, then H is not a finitely generated group.

To see this argue the contrapositive as follows. The kernel H acts freely on the discrete set of vertices $f^{-1}(t)$ with quotient equal to the single vertex which is the l preimage of the base point of S^1. Thus, we have a 1-1 correspondence between the vertices of $f^{-1}(t)$ and the elements of H. If H were finitely generated, we could attach finitely many H-equivariant families of 1-cells to this set of vertices to obtain a connected Cayley graph of H. This would in turn imply that $H_0(f^{-1}(t); \mathbb{Z})$ is finitely generated as a $\mathbb{Z}H$-module (by the 0-spheres corresponding to the endpoints of the edges of the Cayley graph).

Morse theory proof of the topological properties of the level set.

We shall show that it is impossible to attach finitely many $\mathrm{Ker}(F_2 \to \mathbb{Z})$-equivariant classes of 1-cells to $f^{-1}(t)$ to obtain a connected 1-complex. Thus $\mathrm{Ker}(F_2 \to \mathbb{Z})$ is not finitely generated.

We argue by contradiction. Suppose there are finitely many 0-spheres $z_1, \ldots, z_k$ in $f^{-1}(t)$ with the property that if we attach k $\mathrm{Ker}(F_2 \to \mathbb{Z})$-equivariant families of 1-cells to $f^{-1}(t)$ using these k 0-spheres as representatives, then $f^{-1}(t)$ becomes connected. Do this in the contractible ambient complex T. Each 0-sphere z_j is the boundary of an embedded interval I_j in the tree T. If this interval has combinatorial length l_j, then it must be contained in the preimage $f^{-1}([t - l_j/2, t + l_j/2])$. Thus we have connected up all the path components of $f^{-1}(t)$ inside the preimage $f^{-1}([t - L, t + L])$ where $L = \max\{l_j/2 \mid 1 \leqslant j \leqslant k\}$.

But the Morse Lemma states that in expanding the preimages from $f^{-1}(t)$ to $f^{-1}([t - L, t + L])$ the only changes in topology are homotopy equivalences or coning off 0-spheres. In particular, no new path components are created. Thus

$f^{-1}([t-L, t+L])$ is path connected. This is denoted by the shaded region in the schematic diagram in Figure 2.6.

But this gives an immediate contradiction. For example, consider a vertex $v \in T$ at height $t+L+1$. There are two distinct edges connecting v to two distinct vertices of $f^{-1}([t-L, t+L])$. Since the latter complex is path connected, this implies that there is a nontrivial 1-cycle in the tree T. See Figure 2.6.

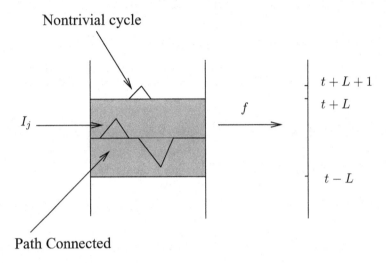

Figure 2.6: Schematic of the contradiction in the proof that $\mathrm{Ker}(F_2 \to \mathbb{Z})$ is not finitely generated.

Remark 2.4.3. Note that the proof above intuitively says that there should be independent generators of $\mathrm{Ker}(F_2 \to \mathbb{Z})$ corresponding to each height above or below level t. A vertex at height n above $f^{-1}(t)$ will correspond to something like a generator $x^n y^{-n}$ for the kernel. The reader should think about how to make this precise, and build a dictionary between the algebra of a generating set for the kernel and the geometry of level sets.

2.4.2   A non-finitely presented example: $\mathrm{Ker}(F_2 \times F_2 \to \mathbb{Z})$

Here we show that the kernel of the map $\varphi : F_2 \times F_2 \to \mathbb{Z}$ which takes all 4 generators of the free groups to a generator of $\mathbb{Z}$ is finitely generated but not finitely presentable.

As before we have a topological model. Let $S^1 \vee S^1$ have one 0-cell and two 1-cells as above. Now give the product $(S^1 \vee S^1) \times (S^1 \vee S^1)$ the product cell structure: one 0-cell, four 1-cells, and four square 2-cells. There is a continuous map $l : (S^1 \vee S^1) \times (S^1 \vee S^1) \to S^1$ which maps the 0-cell to the 0-cell, each open 1-cell homeomorphically onto the open 1-cell in the image, and extends "linearly" over the square 2-cells. Check that l_* is the map φ above. The map l lifts to a φ-equivariant Morse function on the universal cover $T \times T$ where T is the infinite 4-valent tree. Note that this cover is a contractible 2-complex.

Again, we pick an integer point $t \in \mathbb{R}$ and look at the topology of the level set $f^{-1}(t)$. This is now a graph in the 2-complex $T \times T$ (composed of diagonals of the square 2-cells which intersect the level $f^{-1}(t)$). As before, $H = \mathrm{Ker}(F_2 \times F_2 \to \mathbb{Z})$ acts on this graph by restriction of deck transformations with quotient equal to the compact preimage of the base point of S^1 under the map l.

Reduction to topological properties of level sets.

The two algebraic properties of the kernel H derive from topological properties of $f^{-1}(t)$.

1. If $f^{-1}(t)$ is connected, then H is finitely generated.

2. If $H_1(f^{-1}(t); \mathbb{Z})$ is not finitely generated as a $\mathbb{Z}H$-module, then H is not finitely presentable.

The first implication follows from a fundamental result in covering space theory. If $\rho : X \to Y$ is a regular covering space with deck group H, then we have a short exact sequence

$$1 \to \pi_1(X) \to \pi_1(Y) \to H \to 1$$

where the homomorphism $\pi_1(X) \to \pi_1(Y)$ is ρ_*. In our case $X = f^{-1}(t)$ and Y is the compact preimage of the base point of S^1 under the map l. In particular $\pi_1(Y)$ is a finitely generated free group, and so H is also finitely generated.

To see the second implication we can argue the contrapositive statement as follows (assuming that we already know $f^{-1}(t)$ is connected). The quotient Y is a bouquet of 4 circles, which represent the generators of H. The vertices of the graph $f^{-1}(t)$ are in 1-1 correspondence with the elements of H, and the edges correspond to the generators. Thus, $f^{-1}(t)$ is a Cayley graph of H.

If H were finitely presented, we could attach finitely many relator 2-cells to Y to get a presentation 2-complex for H. Attaching finitely many H-equivariant families of lifts of these 2-cells to $f^{-1}(t)$ would result in a 1-connected Cayley complex for H. Thus $H_1(f^{-1}(t); \mathbb{Z})$ is finitely generated as a $\mathbb{Z}H$-module (by the images of the attaching maps of lifts of the relator 2-cells).

Morse theory proof of the topological properties of the level sets.

We start with a local analysis of the complex $(S^1 \vee S^1) \times (S^1 \vee S^1)$. The link of the vertex in $S^1 \vee S^1$ is a set of 4 points; two of which form the ascending link and the other two form the descending link. On taking the product of two copies we get a 2-complex with one vertex whose link is the non-planar graph obtained by forming the join of two sets of 4 points. There are two circles contained in this graph; one is the ascending link and the other is the descending link (they are joins of two ascending/descending 0-spheres, one in each factor).

What does this tell us about the neighborhood of a typical vertex in $T \times T$ from the perspective of the Morse function f? Recall that the restriction of f to any square 2-cell has a maximum at a vertex, a minimum at a vertex, and the other two vertices are on the same mid-level. There are $4 \times 4 = 16$ edges in the link, and so there are 16 square 2-cells adjacent to v in $T \times T$. There is an $S^0 * S^0$ as the ascending link and an $S^0 * S^0$ as the descending link. This means that there are 4 square 2-cells adjacent to, but above, v in $T \times T$. They form a piece of a plane with v as the center vertex. Likewise there is a planar configuration of 4 square 2-cells below v with v as the center. The remaining 8 square 2-cells adjacent to v have v as a mid-level vertex.

The Morse Lemma tells us that when we build up larger and larger interval preimages from a level set $f^{-1}(t)$, that the only changes in topology are homotopy equivalences and coning off circles (in the form of $S^0 * S^0$). (You should try to prove this fact directly using the local model of the preceding paragraph). In particular, neither operation of homotopy equivalence or coning off a circle affects connectivity. Since the space $f^{-1}((-\infty, \infty)) = T \times T$ is path connected, we conclude that $f^{-1}(t)$ is path connected. This is topological property 1. So H is finitely generated.

We argue that $H_1(f^{-1}(t); \mathbb{Z})$ is not finitely generated as a $\mathbb{Z}H$-module by contradiction. If it were, it would be possible to kill H_1 by killing off all H-translates of a finite collection $z_1, \ldots, z_k$ of 1-cycles. We can do this inside the contractible 2-complex $T \times T$. Each z_j bounds a compact 2-chain in $T \times T$. By compactness and finiteness, these 2-chains are contained in $f^{-1}([t - L, t + L])$ for some integer $L > 0$. Since H acts "horizontally" all H-translates of these 2-chains lie in $f^{-1}([t - L, t + L])$ too. This means that the inclusion induced map $H_1(f^{-1}(t); \mathbb{Z}) \to H_1(f^{-1}([t - L, t + L]); \mathbb{Z})$ is the 0 map. However, by the Morse Lemma, $f^{-1}([t - L, t + L])$ is obtained from $f^{-1}(t)$ by homotopy equivalences and coning off circles. Neither operation creates new H_1, and so the inclusion induced map above is also an epimorphism. Thus, $H_1(f^{-1}([t - L, t + L]); \mathbb{Z}) = 0$.

But again we get an immediate contradiction. If v is a vertex at height $t + L + 1$. The union of the "horizontal" diagonals of the 4 squares in $T \times T$ which correspond to the descending link of v is a circle which is contained in $f^{-1}([t - L, t + L])$. Since $H_1(f^{-1}([t - L, t + L]); \mathbb{Z}) = 0$ this circle bounds a 2-chain in $f^{-1}([t - L, t + L])$. This 2-chain together with the cone to v gives a nontrivial 2-cycle in the contractible 2-complex $T \times T$. This is a contradiction.

Remark 2.4.4 (Schematic Figure 2.6 works in all dimensions). We can use the schematic of Figure 2.6 with the proof above. Now interpret the left-hand side as the contractible 2-complex $T \times T$ (instead of the contractible 1-complex T). The shaded region is the preimage $f^{-1}([t - L, t + L])$ in both examples, but now it is homologically 1-connected (instead of just path connected). The z_j are now 1-cycles, and the cycle based at the vertex v is now a nontrivial 2-cycle. Formally, the proofs are identical.

One can think of the level set in this case as the Cayley graph of the group H. The non finite presentability of H corresponds to vertices of arbitrary height above the level set $f^{-1}(t)$. Write down explicit relator words in the generators of H corresponding to vertices v of height n above $f^{-1}(t)$. There are very natural choices which have length $4n$.

2.4.3 A non-F_3 example: $\mathrm{Ker}(F_2 \times F_2 \times F_2 \to \mathbb{Z})$

We show that the kernel H of the map $\varphi : F_2 \times F_2 \times F_2 \to \mathbb{Z}$ which maps all 6 generators of the free factors to a generator of $\mathbb{Z}$ is finitely presented but not of type F_3. The group H was first introduced by Stallings in [36], and was first seen to be a subgroup of a direct product of free groups in [5].

The topological representative is a map l from the direct product of three copies of the complex $S^1 \vee S^1$ of the first example. The product cell structure consists of one 0-cell, six 1-cells, twelve square 2-cells, and eight cubical 3-cells. The map l to S^1 is a bijection of the 0-skeleta, takes each of the six open 1-cells homeomorphically onto the open 1-cell in S^1, and extends "linearly" over the higher cells. The map on π_1 is φ. It lifts to a Morse function f on the universal cover $T \times T \times T$.

The link of the single vertex in the 3-complex is the 3-fold join of a set of 4 points. The ascending and descending links are 3-fold joins of 0-spheres. This join of 0-spheres is a 2-sphere, triangulated as the boundary of an octahedron. Take an integer $t \in \mathbb{R}$. The kernel H acts by restriction of deck transformations on the level set $f^{-1}(t)$ with compact quotient. This quotient is equal to the l preimage of the base point of S^1. It is a nice exercise to show that this preimage is just the 2-complex $T(K)$ obtained by applying the torus construction to the 2-sphere $K = S^0 * S^0 * S^0$. See Figure 2.7 for a hint.

Reduction to topological properties of level sets.

The two algebraic properties of the kernel H derive from topological properties of $f^{-1}(t)$.

1. If $f^{-1}(t)$ is 1-connected, then H is finitely presented.

2. If $H_2(f^{-1}(t); \mathbb{Z})$ is not finitely generated as a $\mathbb{Z}H$-module, then H is not of type F_3.

The first implication is trivial, since then $f^{-1}(t)$ is the universal cover of the compact quotient complex. Thus H is the fundamental group of this finite complex, and so is finitely presented.

We do not give a "hands on" topological proof of the second implication. Instead we appeal to Schanuel's Lemma, [19], from homological algebra which takes the hypothesis that H is of type F_3 and concludes that $H_2(f^{-1}(t); \mathbb{Z})$ is finitely generated as a $\mathbb{Z}H$-module.

Morse theory proof of the topological properties of the level sets.

The Morse Lemma implies that the only changes in topology when taking the f preimages of larger and larger intervals are homotopy equivalences or coning off 2-spheres. Neither operation affects 1-connectedness. Since the whole complex $f^{-1}((-\infty, \infty))$ is contractible, we conclude that $f^{-1}(t)$ is 1-connected.

We argue the second implication by contradiction. Follow the outline in the previous two examples, only use H_2 in place of H_1 or path connectedness. One should obtain an interval $[t-L, t+L]$ whose preimage is homologically 2-connected. Looking at the descending link of a vertex v on level $t+L+1$ gives a way of producing a nontrivial 3-cycle in the contractible 3-complex $T \times T \times T$, a contradiction. As we already stated in Remark 2.4.4, the schematic diagram in Figure 2.6 can be used to guide you through the formal proof.

2.4.4 Branched cover example

The main example of Section 2.6 concerns a compact 3-dimensional cubical complex with hyperbolic fundamental group which is obtained as a branched cover of some simpler complex.

The key features of this example from our current perspective is that its universal cover is CAT(0) and so contractible, that it has a circle-valued Morse function which defines an epimorphism φ from its fundamental group to $\mathbb{Z}$. The Morse function lifts to a φ-equivariant Morse function on the universal cover. The ascending and descending links of vertices are all 2-spheres.

Since the 2-sphere is 1-connected, we conclude that the level sets are 1-connected, and hence that $\mathrm{Ker}(\varphi)$ is finitely presented. Since $H_2(S^2; \mathbb{Z}) \neq 0$ we conclude that and the ambient cubical complex is 3-dimensional we conclude that $\mathrm{Ker}(\varphi)$ is not of type F_3.

Since the ambient group has a 3-dimensional $K(G, 1)$ space, we conclude that it is torsion free. Hence $\mathrm{Ker}(\varphi)$ is also torsion free. We conclude that $\mathrm{Ker}(\varphi)$ could not be a hyperbolic group. If it were hyperbolic, it would be a torsion free hyperbolic group, and so by the second example following Definition 2.4.1 it would be of type F_n for all $n \geqslant 1$. This contradicts the fact that it is not of type F_3.

It is worth pointing out that we used topological information (failure of the finiteness property F_3) to conclude that the kernel above is not hyperbolic. So it should have a Dehn function which is at least quadratic. By Hamish Short's part of the course, we know that the Dehn function of this kernel is bounded above by a polynomial function. We still do not know the exact Dehn function of this group.

2.5 Right-angled Artin group examples

In the first subsection below we define right-angled Artin groups, and their associated non-positively curved cubical Eilenberg–Mac Lane spaces. In the second

subsection we look closely at the geometry of certain kernel subgroups of a particular family of 3-dimensional right-angled Artin groups. This is our second family
of CAT(0) groups containing subgroups with Dehn functions an arbitrary degree
polynomial. Unlike the examples arising from the Bieri doubling trick, the Dehn
functions of these groups do not appear to arise from highly distorted subgroups.

There are interesting directions in which to develop these examples. One
could ask about the Dehn functions of kernel subgroups of other right-angled Artin
groups, or more general CAT(0) cubical groups. There is very little known here; for
example, the precise Dehn function of the Stallings' group, $\mathrm{Ker}(F_2^3 \to \mathbb{Z})$, is not
known at the time of this writing. In another direction, one could investigate the
higher order Dehn functions of subgroups of right-angled Artin groups. Nothing is
known here. Finally, one could try to combine different techniques for producing
CAT(0) groups which contain subgroups with high Dehn functions to produce new
examples. As of this writing, there are only examples of CAT(0) groups containing
subgroups with arbitrary degree polynomial Dehn functions or with exponential
Dehn functions. The methods either involve the right-angled Artin trick described
below, or they use highly distorted subgroups to achieve high Dehn functions
(as is the case with Bieri doubling; exponential Dehn function examples can be
manufactured this way too). Currently there are no known examples of CAT(0)
groups which contain subgroups with Dehn functions distinct from polynomial
or exponential. It would be very interesting to find such examples; for instance,
Dehn functions of the form x^α for some $\alpha \in [2, \infty) \setminus \mathbb{Z}$ or to find super exponential
examples.

2.5.1 Right-angled Artin groups, cubical complexes and Morse theory

Definition 2.5.1 (Right-angled Artin group). Let Γ be a finite graph with n vertices
$\{v_1, \ldots, v_n\}$. The *right angled Artin group* A_Γ associated to Γ has the presentation

$$A_\Gamma = \langle a_1, \ldots, a_n \mid [a_i, a_j] = 1 \text{ whenever } v_i \text{ and } v_j \text{ are adjacent in } \Gamma \rangle.$$

Definition 2.5.2 (Cubical complex associated to a right-angled Artin group). Given
a finite graph Γ, let K_Γ denote the finite flag complex determined by Γ (i.e. the
unique flag complex which has Γ as its 1-skeleton). If Γ has n vertices, we can
realize each of these as the endpoints of standard unit basis vectors in $\mathbb{R}^n$, and by
taking convex hulls of various sets of vertices in $Lk(\mathbf{0}, \mathbb{R}^n)$, we obtain a geometric
realization of K_Γ in the unit $(n-1)$-sphere. Now let C_Γ denote the union of unit
cubes in $\mathbb{R}^n$ based at the origin $\mathbf{0}$ so that $Lk(\mathbf{0}, C_\Gamma)$ is precisely this geometric
realization of K_Γ. Let X_Γ denote the quotient of C_Γ by the equivalence relation
defined by $\mathbf{x} \sim \mathbf{y}$ if $\mathbf{x} - \mathbf{y} \in \mathbb{Z}^n$. Thus, X_Γ is obtained from the union of cubes
C_Γ by identifying opposite faces by integer translations. Note that X_Γ is just a
collection of tori of various dimensions glued together by isometries along standard
(coordinate) subtori. Check the following.

1. X_Γ has one vertex. Call it v.

2. X_Γ has n 1-cells.

3. X_Γ has one square 2-cell for each edge of Γ.

4. $\pi_1(X_\Gamma) = A_\Gamma$. Why is it needed to look beyond the 2-skeleton of X_Γ in your determination of π_1?

5. $Lk(v, X_\Gamma) = S(K_\Gamma)$ where $S(.)$ denotes the spherical construction from Definition 2.1.22.

6. Conclude that X_Γ is a non-positively curved PE cubical complex.

Remark 2.5.3 (2-complex case). Suppose that the flag complex corresponding to the graph Γ is a 2-dimensional simplicial complex K_Γ such that each edge and each vertex is contained in at least one 2-simplex. Then each 2-simplex of K_Γ determines a 3-torus in the corresponding right-angled Artin cubical complex. The preimage of a basepoint under the circle-valued Morse function will be precisely the torus complex $T(K_\Gamma)$ of Definition 1.2.7. Figure 2.7 gives a picture of a 2-simplex and its companion simplex inside a 3-cube. Remember that the opposite faces of the 3-cube are identified by translations to give a 3-torus in the right-angled Artin cubical complex. In fact, this situation was the inspiration for the definition of the torus construction.

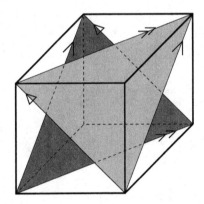

Figure 2.7: A triangle and its companion in a 3-torus of a right-angled Artin cubical complex.

Remark 2.5.4 (Notation). We shall switch back and forth between a finite flag complex K and its underlying graph $K^{(1)}$. When referring to the corresponding right-angled Artin group, cubical complex etc we shall often use the notation A_K, X_K etc instead of the original notation $A_{K^{(1)}}$, $X_{K^{(1)}}$, etc. which refers explicitly to the underlying graph.

So given a flag 2-complex K we obtain a right-angled Artin cubical 3-complex, denoted by X_K, and a torus complex $T(K)$ which is the preimage of the basepoint under the circle-valued Morse function. Does $T(K)$ π_1-inject into X_K? Equivalently, are the lifts of $T(K)$ to the universal cover of X_K all copies of the universal cover of $T(K)$? These lifts are precisely the level sets $f^{-1}(t)$ of the lift of the Morse function to the universal covers of X_K and of the the target circle. The next theorem from [4] gives detailed information about the topology of these level sets $f^{-1}(t)$. The answer only depends on the topology of the original complex K.

Theorem 2.5.5 (Level set topology). *Let K be a finite flag complex, and X_K be the corresponding right-angled Artin cubical complex. Let $f : \widetilde{X_K} \to \mathbb{R}$ be the lift of the circle valued Morse function on X_K. Then $f^{-1}(t)$ is homotopy equivalent to an infinite wedge of copies of K, one for each vertex of $\widetilde{X_K}$ which is not contained in $f^{-1}(t)$.*

This answers our question about π_1-injectivity of $T(K)$ into X_K. If K is 1-connected, then $f^{-1}(t)$ is also 1-connected. For integer values of t, this is a cover of the torus complex $T(K) \subset X_K$. In the next subsection, the flag complexes K_i are all triangulations of the disk D^2. Thus the corresponding torus complexes $T(K_i)$ all π_1-inject into X_{K_i}, and the kernel subgroup H_{K_i} is just $\pi_1(T(K_i))$. The integer level sets in $\widetilde{X_{K_i}}$ are contractible universal covers of $T(K_i)$.

Example (Right-angled Artin examples). Fill in the table in Figure 2.15 at the end of this chapter. K is a flag complex, $S(K)$ the associated spherical complex, $T(K)$ the torus construction (in the case K is 2-dimensional), X_K the cubical complex X_K, A_K the Artin group, and H_K the kernel. Some entries in the table have already been filled in as a guide.

The next example shows that we could have taken the vertex groups V_m in the graph of groups description of the snowflake groups to be right-angled Artin groups.

Remark 2.5.6 (Vertex groups revisited). Verify that the vertex group V_m of Definition 1.2.8 is isomorphic to the right-angled Artin group A_K where K is a segment of length $m - 1$. Label the m generators of A_K as $u_1, \ldots, u_m$, where u_1 and u_m correspond to the end vertices of K. Verify that one can define an isomorphism by mapping the generator a_1 of V_m to u_1, mapping a_i to $u_{i-1}^{-1} u_i$ for $i \geqslant 2$, and mapping c to u_m.

2.5.2   The polynomial Dehn function examples

We show that the right-angled Artin groups based on the sequence $\{K_n\}$ of flag complexes shown in Figure 2.8 below have kernel subgroups whose Dehn functions are of the form x^{n+2}. These examples are from joint work of the author and Max Forester and Krishnan Shankar. The kernel subgroups are the kernels of the map from the right-angled Artin presentation to $\mathbb{Z}$, taking each generator to a generator of $\mathbb{Z}$. From the Morse theory results above, we know that the kernels are the fundamental groups of the torus complexes $T(K_n)$, and also that the $T(K_n)$ are aspherical 2-complexes. Their contractible universal covers are precisely the integer preimage level sets of the Morse function from the universal cover of the Artin cubical complex to $\mathbb{R}$.

We argue that the Dehn function of the groups $H_{K_n} = \pi_1(T(K_n))$ are x^{n+2} by giving lower and upper bounds of this form. Here is an overview.

The lower bounds are established by defining a sequence of loops $\{w_k\}_{k=1}^{\infty}$ in $\pi_1(T(K_n))$ whose area is at least $|w_k|^{n+2}$. The area of each loop is computed by constructing an embedded van Kampen diagram for it in the universal cover of $T(K_n)$. Since this universal cover is a contractible 2-complex, we conclude (just as in the case of the snowflake groups) that the area of the loop is equal to the area of the embedded diagram.

The upper bounds are established by induction on n. The key idea is the same as in the upper bound proof for the subgroups of $F_2 \times (F_n \rtimes_\varphi \mathbb{Z})$ from Section 2.3. We cut arbitrary van Kampen diagrams over $\pi_1(T(K_n))$ along certain corridors to get a union of corridors and complimentary van Kampen diagrams over $\pi_1(T(K_{n-1}))$. This allows the induction to work. One needs to check that the boundary words of the corridors are undistorted free subgroups of $\pi_1(T(K_n))$.

There are more steps listed in the lower and upper bound arguments below. However, they still just suggest how the arguments should proceed in each case. It is a good exercise to fill in the details of these steps and give a rigorous argument in each case. Details can be found in [11].

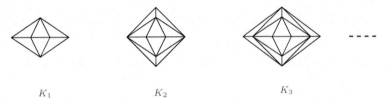

K_1 $\qquad\qquad$ K_2 $\qquad\qquad$ K_3

Figure 2.8: Right-angled Artin groups containing subgroups whose Dehn function is similar to x^{n+2}.

Lower Bounds. We show that x^{n+2} is a lower bound for the Dehn function of $\pi_1(T(K_n))$ by exhibiting van Kampen diagrams with boundary length a linear function of x and whose area is of order x^{n+2}, and which embed in the contractible universal covers of $T(K_n)$. Since the universal cover of $T(K_n)$ is 2-dimensional and contractible, the embedded van Kampen diagrams give a lower bound for the area of their boundary loops.

1. Start with the flag complex K_1. Label the 4 boundary edges of the complex K_1 as follows. Label the oriented edge from the leftmost (resp. rightmost) vertex on the middle row to the top vertex by a (resp. b). Label the oriented

edge from the leftmost (resp. rightmost) vertex on the middle row to the bottom vertex by p (resp. q). Consider the van Kampen diagram for the word w_k defined as

$$(ab)^k (a^{-1}b^{-1})^k (p^{-1}q^{-1})^{-k} (pq)^{-k}.$$

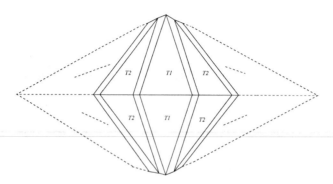

Figure 2.9: Cubic area van Kampen diagram for $\pi_1(T(K_1))$.

Check that it has area of order k^3, and so order $|w_k|^3$. The four 2-tori which contain the generators a, b, p and q will contribute only linear area to this van Kampen diagram. These correspond to the strips (corridors) in Figure 2.9. The inner two triangles of K_1 give two 2-tori in $T(K_1)$ which contribute most of the area. The area of the T_1 triangles in Figure 2.9 is k^2, the area of the T_2 triangles is $(k-1)^2$, and so on. The total contribution to the area by these triangles is bounden below by $\sum_{i=1}^{k} i^2$ which is of order k^3.

2. Verify that the diagram above does indeed embed in the universal cover of $T(K_1)$. Conclude that there is a lower bound of x^3 for the Dehn function of $\pi_1(T(K_1))$.

3. Now proceed in a similar fashion with K_2. It will help to turn K_2 on its side (rotate by $\pi/2$) before labeling the boundary 4 edges as above. Use similar words to the words in the previous step. This time however, the "inner diamond" of two T_1 triangles will be replaced by the van Kampen diagram from step 1 above. Likewise for the diamond of T_2 triangles etc. Thus one should obtain a van Kampen diagram with boundary length a linear function of k, and area bounded below by $\sum_{i=1}^{k} i^3$ which is of order k^4. It is worth writing out the words w_k explicitly in this case, and seeing how the van Kampen diagram is built up from previous van Kampen diagrams over $\pi_1(T(K_2))$.

4. Repeat the argument inductively for K_3 (do not rotate), K_4 (rotate by $\pi/2$), and so on. Write down a general expression for the sequence $\{w_k^{(n)}\}_{k=1}^{\infty}$ of words in $\pi_1(T(K_n))$ for each $n \geqslant 1$.

Upper Bounds. Let K_0 denote the suspension of a line segment; that is, K_0 is a 2-complex consisting of a pair of 2-simplices attached along a common edge. Note that K_1 is obtained from K_0 by attaching two other copies of K_0, each attached along a pair of edges. In general, K_{n+1} is obtained from K_n by attaching two copies of K_0, each attached along a pair of edges.

The torus complex $T(K_0)$ is just a pair of 2-tori glued together along a common circle. The fundamental group is $F_2 \times \mathbb{Z}$ which has quadratic Dehn function. So the main point of this section is to argue that the Dehn functions of the torus complexes that arise after each coning process increases by at most one power in x.

We need a torus complex description of what happens when we attach two copies of K_0 to K_n, with the attaching along pairs of adjacent edges. From the perspective of torus complexes, we are attaching two copies of $T(K_0)$, each glued to $T(K_n)$ along a bouquet of two circles. It is a good exercise to see that $T(K_{n+1})$ retracts onto each copy of these bouquets of two circles. (Hint: by functorality of the torus construction, it suffices to check that each K_{n+1} admits simplicial retractions onto each of the two pairs of edges which form half of the original boundary of the K_n subcomplex. This is easy to see inductively). Therefore $T(K_{n+1})$ admits retractions onto each of the two bouquets of two circles, along which the copies of $T(K_0)$ attach to $T(K_n)$ to yield $T(K_{n+1})$.

Since $\pi_1(T(K_0)) = F_2 \times \mathbb{Z}$, we obtain the following decomposition of fundamental groups:

$$\pi_1(T(K_{n+1})) = (F_2 \times \mathbb{Z}) *_{F_2} \pi_1(T(K_n)) *_{F_2} (F_2 \times \mathbb{Z})$$

where the amalgamating F_2 are retracts of $\pi_1(T(K_{n+1}))$. In particular, these F_2 groups are undistorted in $\pi_1(T(K_{n+1}))$. Now finish these steps.

1. Analyze van Kampen diagrams over $\pi_1(T(K_{n+1}))$ from the perspective of the amalgam above. Think of the $\mathbb{Z}$ factors in the $F_2 \times \mathbb{Z}$ as determining corridors in van Kampen diagrams over the generators of $\pi_1(T(K_{n+1}))$.

   These corridors do not form annuli in a reduced van Kampen diagram of a word w (why?). Therefore these corridors begin and end on the boundary word w of the diagram.

2. The lengths of these corridors is at most the length of w (why?).

3. The corridors chop the original diagram into at most $|w|/2$ strips and at most $|w|/2$ complimentary regions. These complimentary regions are van Kampen diagrams over the generators of $\pi_1(T(K_n))$ with boundary lengths bounded above by $|w|$.

4. Finish the upper bound proof by induction on n.

2.6   A hyperbolic example

We have seen several examples in the previous sections which demonstrate that subgroups of CAT(0) groups can have polynomial Dehn functions of arbitrary degree. In this section we explore the question of whether something analogous is true for subgroups of hyperbolic groups. The situation is far less understood. At present there is only one construction [8] which gives examples of hyperbolic groups containing finitely presented subgroups which are not hyperbolic. The subgroups are known to be not hyperbolic because, like the Stallings group, they are not of type F_3. We do not have explicit calculations of their Dehn functions, although they are known to be bounded above by polynomials [28].

The examples arise as kernels of maps from the fundamental groups of non-positively curved cubical 3-complexes to $\mathbb{Z}$. The non-positively curved 3-complexes are obtained as *branched covers* of direct products of finite graphs.

This section is organized as follows. We begin by extending the manifold notions of *codimension 2 branching locus* and of *branched cover* to the realms of non-positively curved piecewise Euclidean cubical complexes and of 2-dimensional complexes. We shall see by low dimensional examples, that branched covers of non-positively curved piecewise Euclidean cubical complexes can have hyperbolic fundamental groups. The intuition that one takes from this is that "branched coverings are a way to kill flat tori in non-positively curved complexes", and so are a good way to try to produce hyperbolic groups. The examples begin with different branched covers of the 2-torus, and then increase in complexity. In one example the dimension increases, and we consider branched covers of the 3-torus over a link. In the next example, the local complexity increases, and we consider branched covers of a non-manifold 2-complex over a vertex. Both these last two examples are fundamental for understanding the techniques behind the main construction of this section.

Next we shall see that this intuition about branched covers and hyperbolicity is only a low ($\leqslant 3$) dimensional intuition. In dimension 4 and above, the act of taking branched covers over generic branch loci tends to *produce new flat tori* in addition to killing old flat tori. Thus the branched covers do not have hyperbolic fundamental groups.

Finally, we describe how to construct the examples alluded to in the first paragraph. Since the subgroups should fail to be of type F_3, our starting point is a product of three finite graphs (fundamental group is a product of three free groups). There is a circle-valued Morse function on this cubical 3-complex so that each of the ascending and descending links is a 2-sphere triangulated as the boundary of an octahedron. We choose a branched cover of this cubical 3-complex, so that the fundamental group of the cover is hyperbolic. Moreover, the branch locus and branched cover is chosen so that the ascending and descending links of the induced Morse function on the cover are all 2-spheres. In this example, these 2-spheres are either 3-fold joins of 0-spheres, or are 5-fold branched covers of these 2-complexes over one of the 0-spheres.

2.6.1 Branched covers of complexes

Here are two situations where one can speak about branched covers of complexes over subcomplexes.

Branched covers of 2-complexes.

The first situation involves taking branched covers of 2-complexes over their 0-skeleta. See [26] and [12] for applications. Given a 2-complex, K, with the property that $Lk(v, K)$ is non-empty and connected for all $v \in K^{(0)}$. Now $K \setminus K^{(0)}$ deformation retracts onto a dual graph (this has vertices at barycenters of 2-cells of K, and edges dual to the original 1-cells of K). Note that for each $v \in K^{(0)}$ the link graph $Lk(v, K)$ immerses into the dual (and so π_1-injects) under this deformation retraction. In this situation one can define branched covers by taking suitable covers of the dual graph. These define covers of $K \setminus K^{(0)}$. Think of K as having a piecewise Euclidean metric. Lift the metric to the cover of $K \setminus K^{(0)}$, and then take the completion.

Branched covers of cubical complexes.

The second situation involves branched covers of non-positively curved piecewise Euclidean cubical complexes over suitable branch loci.

Definition 2.6.1 (Cubical branch locus). Let K be a PE cubical complex, and $L \subset K$ a subcomplex. We say that L is a *branching locus* if it satisfies the following two conditions.

1. *Local Convexity:* For each m-cell $\chi_e : \square_e^m \to e$ of K with $e \cap L \neq \emptyset$ we have that $\chi_e^{-1}(L)$ is a disjoint union of faces of $\square_e^m$ (possibly all of $\square_e^m$).

2. *Codimension 2:* $Lk(e, K) \setminus Lk(e, L)$ is nonempty and connected for each cell $e \subset L$.

Remark 2.6.2. The first condition is a local convexity condition for L in K. It is used to describe the neighborhoods of L in K, and to prove that non-positive curvature is preserved on passing to branched coverings.

The second condition is a reformulation of the "codimension 2" condition in branched coverings of manifolds over submanifolds. It ensures that the trivial branched covering of a PE cubical complex K over a branching locus L is just K itself.

Here's what we mean by branched cover in the setting of PE cubical complexes.

Definition 2.6.3 (Cubical branched cover). Let K be a PE cubical complex and $L \subset K$ be a branching locus as above. By a *finite branched covering* of K over L we mean the result of the following process.

1. Take a finite covering of $K \setminus L$.

2. Lift the local piecewise Euclidean metric on $K \setminus L$ to this cover, and then consider the induced path metric.

3. Finally, take the metric completion of this cover.

There are two things that should be true of branched covers of non-positively curved PE cubical complexes; they should be again PE cubical complexes, and they should be non-positively curved. The next two lemmas from [8] say that this is indeed the case.

Lemma 2.6.4. *Let K be a finite PE cubical complex, and let $\hat{K}$ be a finite branched covering. Then $\hat{K}$ is also a finite PE cubical complex, and there is a natural continuous surjective map $b : \hat{K} \to K$. In the case of a trivial branched covering, one has $\hat{K} = K$.*

Lemma 2.6.5. *Let $b : \hat{K} \to K$ be a finite branched covering of a PE cubical complex K of non-positive curvature, and suppose that the branching locus $L \subset K$ is a finite graph. Then $\hat{K}$ is also a non-positively curved PE cubical complex.*

2.6.2   Branched covers and hyperbolicity in low dimensions

Here we explore the idea of using branched covers to remove flat tori from spaces in order to obtain spaces with hyperbolic fundamental group. We proceed by a series of examples, most of which are manifold examples. One is a 2-complex example. Both the 3-dimensional manifold example, and the 2-complex example are important background examples for the construction used in the proof of Theorem 2.6.6 below.

This first example gets you introduced to the idea of branched covers in the manifold setting. It is easy to visualize this branched cover, and to see that flat tori have been "killed" in the branching process.

Example (T^2 over 2 points). Consider the 2-fold branched cover of the 2-torus over 2 points. Start with the product of a circle of length 1 and a circle of length 2. Denote the vertex of the first circle by 0, and the vertices of the second circle by 0 and 1, so that the vertices of the torus are $(0,0)$ and $(0,1)$. The torus has 4 edges and 2 square 2-cells. Construct a 2-fold branched cover of this torus over the two vertices. Note that this branched cover is also a 2-manifold. Compute its genus. Check that it is non-positively curved in the piecewise Euclidean square metric. Does the universal cover contain any embedded flat planes?

The next example is similar in flavor to the first. However it is harder to visualize, as the branching locus in the base space consists of just one point. This example demonstrates that the technique of using representations to describe covering spaces can help greatly in describing more complicated branched covers.

Example (T^2 over 1 point). Construct a 3-fold branched cover of T^2 over one point, where the preimage of the branch point in the base consists of just one point in the cover. The following hints might help.

1. Think of the torus T^2 as being obtained from a square by identifying opposite edges by translations. The cell structure on T^2 consists of one vertex v, two one cells a and b, and a single square 2-cell.

2. Consider the following graph in the square; connect the barycenter of the square to the barycenters of each of the 4 sides. This gives a bouquet of two circles in the torus, which is "dual" to the original cell structure. Check that the torus minus the original vertex v deformation retracts onto this graph. Thus the problem of finding a 3-fold cover of $T^2 \setminus \{v\}$ is reduced to finding a 3-fold cover of this bouquet of two circles. Label the loops in this graph by their images in the fundamental group of T^2. Thus one loop will be labeled a and the other b.

3. Check that the link of v in T^2 gets sent to $[a, b]$ under the deformation retraction above.

4. Recall that 3-fold covering spaces are determined by index three subgroups of the fundamental group of the base space. In this case these subgroups are recorded via representations of $F_{\{a,b\}}$ into the symmetric group on 3 generators. Given a representation $\rho\colon F_{\{a,b\}} \to S_3$ the corresponding index 3 subgroup is defined by

$$H \;=\; \{g \in F_{\{a,b\}} \mid \rho(g)(1) = 1\}.$$

5. We want that the preimage of v consists of only one point. Thus, we require that $Lk(v, T^2)$ has just one preimage in the 3-fold cover. This will happen if the representation is chosen so that $[a, b]$ gets sent to a 3-cycle.

6. Write down a representation $F_{\{a,b\}} \to S_3$ which sends $[a, b]$ to a 3-cycle. Hint: what happens if one of a or b is sent to a 2-cycle and the other is sent to a 3-cycle?

7. Now construct the branched cover of T^2 over $\{v\}$ corresponding to your choice of representation above. This should be a surface tiled by 3 square 2-cells. Describe explicitly the edge pair identifications on this collection of 3 squares. How many edges, vertices, and 2-cells does your branched cover have? It is a closed surface. What is its genus?

8. Again check that the branched cover is non-positively curved in the piecewise Euclidean squared metric. Check also that there are no embedded flat planes in the universal cover of the branched cover.

We move up one dimension with the next example. Again the theme of using branched covers to "kill" flat tori, and so to obtain hyperbolic groups is reiterated in this example. As with the previous example, representations are used to describe the branched cover. This is the first important background example for the construction in the proof of Theorem 2.6.6 below.

Example (T^3 over 3 circles). Here we construct 3-manifold branched covers of the 3-torus over a particular 3 component link.

1. Start with a cubical structure on T^3 given by the product of three circles S^1, each of which has 2 vertices (labeled 0 and 1) and 2 edges. Thus, T^3 has 8 vertices with labels ranging from $(0,0,0)$ through $(1,1,1)$.

2. Take as branching locus the 3 component link

$$L = (S^1 \times \{0\} \times \{1\}) \cup (\{1\} \times S^1 \times \{0\}) \cup (\{0\} \times \{1\} \times S^1).$$

3. Note that projection onto the first two coordinates gives a continuous map

$$p_3 : T^3 \setminus L \to T^2 \setminus \{(0,1)\}.$$

   Thus we can define branched covers of T^3 over the third coordinate component of L by composing p_{3*} with the representation of $\pi_1(T^2 \setminus \{(0,1)\})$ to S_3 given in the previous example. Denote this composition by $\rho_3 : \pi_1(T^3 \setminus L) \to S_3$.

   Similarly, there are representations ρ_1 and ρ_2 obtained by using the projections which kill the first and second coordinates respectively.

4. Now combine these three representations into a single representation $\rho : \pi_1(T^3 \setminus L) \to S_{27}$, the symmetric group on 27 elements as follows. Let $X = \{1,2,3\}$, then for $g \in \pi_1(T^3 \setminus L)$ define the action of $\rho(g)$ via

$$\rho(g)(x_1, x_2, x_3) = (\rho_1(g)x_1, \rho_2(g)x_2, \rho(g)x_3) \quad \text{for } (x_1, x_2, x_3) \in X^3$$

   Take the corresponding branched cover $\widehat{T^3}$ of T^3.

5. Check that the universal cover of $\widehat{T^3}$ does not contain any isometrically embedded flat planes. To see this, argue that a flat plane in the universal cover would develop so as to intersect the lifts of the branch locus transversely, and so would have been branched itself. Readers should take a look ahead at Figure 2.14 if they can not see why this should be so.

   The final example in this section involves a branched cover of a 2-complex over a subset of its 0-skeleton. The idea behind the representation borrows heavily from the discussion in the second example above. However, in this case we need to look at representations into S_5 in order to satisfy all the link conditions. This is the second important background example for the construction in the proof of Theorem 2.6.6.

Example ($\Theta \times \Theta$ over a vertex). Let Θ be the graph with two vertices and 4 edges as in Figure 2.12. Take a branched cover of $\Theta \times \Theta$ over the vertex $(0,0)$ as follows.

1. First check that $\Theta \times \Theta \setminus \{(0,1)\}$ deformation retracts onto the subgraph Δ of $\Theta \times \Theta$ which is spanned by the remaining three vertices. See Figure 2.10 and Figure 2.11 to help.

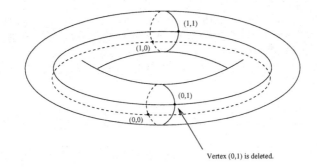

Vertex (0,1) is deleted.

Figure 2.10: A torus in $\Theta \times \Theta \setminus \{(0,1)\}$.

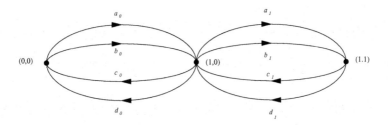

Figure 2.11: The 2-complex $\Theta \times \Theta$ minus the vertex $(0,1)$ deformation retracts onto the graph spanned by the remaining vertices.

2. Check that the collection of 36 loops of combinatorial length 4 in $Lk((0,1), \Theta^2)$ gets mapped bijectively to the collection of commutator loops in the graph Δ. The commutator loops are obtained by choosing a pair of 0-label edges and a pair of 1-label edges, and forming the commutator of the pair of bigon loops based at $(1,0)$.

3. Using the labeling of Figure 2.11, define a homomorphism $\rho : \pi_1(\Delta, (1,0)) \to S_5$ by

$$
\begin{array}{cclcccl}
a_0^{-1}b_0 & \mapsto & \alpha & \qquad & a_1 b_1^{-1} & \mapsto & \beta \\
a_0^{-1}c_0^{-1} & \mapsto & \alpha^2 & \qquad & a_1 c_1 & \mapsto & \beta^2 \\
a_0^{-1}d_0^{-1} & \mapsto & \alpha^3 & \qquad & a_1 d_1 & \mapsto & \beta^2
\end{array}
$$

where $\alpha = (2354)$ and $\beta = (12345)$. Check that for each pair of 0-label edges, and for each pair of 1-label edges, the commutator of the corresponding two (bigon) loops based at $(1,0)$ in Δ are sent to a 5-cycle under ρ. There are 36 pairs to check.

4. Let X denote the corresponding branched cover of $\Theta \times \Theta$ over $(0,1)$. Verify that X is non-positively curved, and that its universal cover does not contain any isometrically embedded flat planes.

5. The edge orientations on Θ define a map to the circle. These oriented edges lift to X, and so we get a map from X^1 to the circle, which extends linearly over the 2-skeleton to give a circle valued Morse function on X. Check that the ascending and descending links of this Morse function are all circles. This property will be crucial in the proof of the main result of this section later on.

6. Argue that the kernel of $X \to S^1$ is finitely generated but not finitely presented. Thus, the group $\pi_1(X)$ is an example of a hyperbolic group which is not coherent. Recall that a group is said to be *coherent* if every finitely generated subgroup is finitely presented.

2.6.3   Branched covers in higher dimensions

The four examples above should have given you the impression that taking branched covers is a good way of producing hyperbolic groups by killing flat tori in the base complexes. This is a good intuition in dimensions 3 or 2. However, it fails in dimension 4 and above for generically chosen branch loci. The problem is that the branching process can "create" new flat planes in addition to "killing" old ones. While considering a branched cover of the 5-torus due to Gromov, Mladen Bestvina noticed that this phenomenon of branched covers "creating" new flat planes can occur in high dimensions. Here we give a 4-dimensional example.

Example (Creating flat planes via branching in dimension 4). Consider what happens when we take a branched cover of the 4-torus over a generically chosen collection of branch 2-tori which are parallel to the coordinate subtori.

There are 6 subtori which we label by their coordinates: (12), (13), (14), (23), (24), (34). To say that they are chosen generically means that (12) and (34) intersect in a point, (13) and (24) intersect in a distinct point, (14) and (23) intersect in a third point, and none of the other pairs intersect.

Note that any flat 2-torus in T^4 will transversely meet at least one of these 6 branch 2-tori, so all the old flat 2-tori will be killed by the branching. Where do the new flat tori arise?

Of the non-intersecting branch tori, choose the pair that are the closest in the usual T^4 metric. Suppose they are (14) and (24). Nearby lifts of these to the universal cover $\mathbb{R}^4$ look like two skew lines in (123)-coordinate space; one of which is parallel to the 1-axis and the other parallel to the 2-axis. There is a common perpendicular segment I to these skew lines. The product of I with the 4th coordinate circle, gives an annulus in T^4 whose boundary circles are contained in the branching loci.

Suppose that the branching degree about (14) is m and the degree about (24) is n. Then the lift of the annulus to the universal cover of the branched cover consists of copies of $\mathbb{R}$ times an bipartite tree with vertex valences m and n. This gives many embedded flat planes in the universal cover of the branched cover.

Note that none of these planes are transversely intersected by any of the other branching loci, since the collection of branch loci is generic, and the initial pair of loci components were chosen to have least distance among all non-intersecting pairs.

2.6.4 The main theorem and the topological version

Here is the main result that we shall discuss for the remainder of the section.

Theorem 2.6.6 (Non-hyperbolic finitely presented subgroup of a hyperbolic group). *There exists a short exact sequence of groups*

$$1 \rightarrow H \rightarrow G \rightarrow \mathbb{Z} \rightarrow 1$$

such that

1. *G is torsion free hyperbolic,*

2. *H is finitely presented, and*

3. *H is not of type F_3.*

In particular, H is not hyperbolic.

Remark 2.6.7. Since H is not hyperbolic, its Dehn function is at least quadratic. Hamish Short has given a sketch of an argument due to himself and Steve Gersten which provides a polynomial upper bound on the Dehn function of H. It would be good to get a precise calculation, or to produce other examples of this sort.

We begin with a purely topological theorem which implies our main theorem.

Theorem 2.6.8 (Topological/local version). *There exists a compact, non-positively curved, piecewise Euclidean cubical complex Y and a continuous map $f : Y \rightarrow S^1$ with the following properties.*

1. *The image of $\varphi = f_* : \pi_1(Y) \rightarrow \pi_1(S^1)$ is of finite index in $\pi_1(S^1)$.*

2. *The universal cover X of Y contains no isometrically embedded flat planes.*

3. *The map f lifts to a φ-equivariant Morse function $\tilde{f} : X \rightarrow \mathbb{R}$ whose ascending and descending links are all homeomorphic to S^2.*

Remark 2.6.9 (Topological/local implies main result). Fill in the steps in the following sketch that the topological/local theorem above implies the main result of this section. You have already encountered this argument in the *Branched Cover Example* of Section 2.4.

1. The universal cover X is CAT(0) since it is non-positively curved and simply connected. In particular, X is contractible. Also, X is a δ-hyperbolic metric space, since it is CAT(0), admits a cocompact group of isometries, and contains no isometrically embedded flat planes. Therefore, G is a hyperbolic

group. Since X is contractible we conclude that Y is a 3-dimensional $K(G, 1)$ space, and therefore that G is torsion free. This establishes conclusion 1 of Theorem 2.6.6.

2. Use the Morse Lemma to conclude that the contractible space X is built up from $f^{-1}(t)$ by homotopy equivalences and coning off 2-spheres. Since neither of these operations affects 1-connectivity, and since the result X is contractible, we conclude that $f^{-1}(t)$ is 1-connected. However, H acts (by restriction of its deck transformation action on all of X) by deck transformations on $f^{-1}(t)$ with compact quotient. Thus H is finitely presented. This establishes point 2 of Theorem 2.6.6.

3. We argue that H is of type F_3 by contradiction. By Schanuel's Lemma from homological algebra, if H were of type F_3, then $H_2(f^{-1}(t); \mathbb{Z})$ would be finitely generated as a $\mathbb{Z}H$-module. This means that there is a finite collection of 2-cycles

$$z_1, \ldots, z_k$$

whose H translates generate all of $H_2(f^{-1}(t), \mathbb{Z})$.

Now each 2-cycle z_i bounds a 3-chain c_i in the contractible 3-complex X. By compactness of 3-chains, the collection of c_i lie in $f^{-1}([t - N, t + N])$ for some $N > 0$. So do the H-translates of the c_i. Therefore, the inclusion induced map $H_2(f^{-1}(t); \mathbb{Z}) \to H_2(f^{-1}([t - N, t + N]); \mathbb{Z})$ is the 0-map.

But the Morse Lemma implies that $f^{-1}([t - N, t + N])$ is obtained from $f^{-1}(t)$ by homotopy equivalences and coning off 2-spheres. In particular, the inclusion induced map on H_2 is an epimorphism (use Mayer–Vietoris). But if this map is onto and is the 0-map, we have that $H_2(f^{-1}([t-N, t+N]); \mathbb{Z}) = 0$.

Now consider a vertex v such that $t + N < f(v) < t + N + 1$ (choose N so that $N + t \notin \mathbb{Z}$). The descending link of v in X is a 2-sphere. There is a geometric realization, S, of this 2-sphere in $f^{-1}(t + N)$. But S binds a 3-chain in $f^{-1}([t - N, t + N])$ since $H_2 = 0$. Also S bounds a 3-chain which is just the cone on S with cone point v. The union of these two 3-chains (with appropriate signs) gives a 3-cycle in a contractible 3-complex, a contradiction.

2.6.5 The main theorem: sketch

The base 3-complex, Morse function, and the branch loci

Let Θ denote the graph shown in Figure 2.12. The base complex for the branched cover construction is Θ^3. Fill in the details of the following steps. The paper [8] gives details.

1. Check that this is a non-positively curved cubical complex. How many vertices, edges, 2-cells and 3-cells does it have?

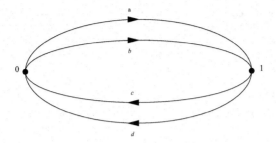

Figure 2.12: The graph Θ.

2. Map each edge of Θ^3 to a circle by sending both end vertices to the basepoint of the circle, and mapping each edge around the oriented circle as specified by the arrow on the edge in Θ^3. This extends linearly to a circle valued Morse function on Θ^3. Describe the ascending and descending links of each vertex of Θ^3.

3. Use Morse theory to argue that $\mathrm{Ker}(\pi_1(\Theta^3) \to \pi_1(S^1))$ is finitely presented, but not of type F_3. Argue exactly as in Remark 2.6.9.

4. Consider the following branch locus:

$$L = (0 \times 1 \times \Theta) \cup (\Theta \times 0 \times 1) \cup (1 \times \Theta \times 0).$$

Check that this is indeed a branch locus according to our definition, and therefore that branched covers of Θ^3 over L are non-positively curved PE cubical complexes.

The branched cover via coordinate projections

We take the branched cover of Θ^3 over L similar to the way we took the branched cover of the 3-torus in an earlier example.

1. There are three projections of $\Theta^3 \setminus L$ onto Θ^2 minus a vertex; each is obtained by killing one of the three coordinate directions. The target 2-complex deformation retracts onto the full subcomplex of its one skeleton which contains the remaining three vertices. Figure 2.11 contains a copy of this in the case that the vertex $(0, 1)$ is removed from Θ^2. Thus, we can compose the projection induced homomorphism with the representation of $\pi_1(\Theta^2 \setminus \{(0, 1)\})$ to S_5 from the earlier example in this section. This gives three representations $\{\rho_i\}_{i=1}^3$ of $\pi_1(\Theta^3 \setminus L)$ to S_5.

2. Define a 125-fold branched cover, denoted by Y, of Θ^3 over L by combining the three representations to S_5 in the same way as in the branched cover of the 3-torus example.

Properties of the finite cover

In this section we show that the branched cover Y above, together with the circle-valued Morse function obtained by composing the branched covering map with the map $\Theta^3 \to S^1$, satisfy the conditions of Theorem 2.6.8.

1. We already know that Y is a compact, non-positively curved, piecewise Euclidean cubical complex. Also, it is easy to check that the circle-valued Morse function induces a homomorphism onto a finite index subgroup of $\mathbb{Z}$.

2. The universal cover of Y contains no embedded flat planes. The idea is to show that a flat plane must intersect the branch locus transversely, giving a contradiction.

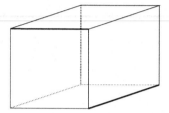

Figure 2.13: The branch locus intersection pattern with a typical 3-cell.

   Figure 2.13 shows how the branch locus intersects a typical cube. It is possible to develop a portion of a flat plane in a single 3-cube which misses the branch locus. However, if we try to develop it into an $\mathbb{R}^3$ neighborhood of a vertex which does not intersect the branching locus, we find that it must hit some component of the branching locus transversely. Why? See Figure 2.14.

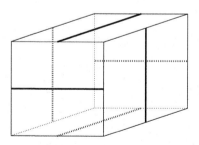

Figure 2.14: A Euclidean neighborhood of a vertex which misses the branch locus.

3. Finally, check that the ascending and descending links of the Morse function are all 2-spheres. There will be two combinatorial types. The first is just the

join $S^0 * S^0 * S^0$ of three 0-spheres. The second is a 5-fold branched cover of this 2-complex over one of the 0-spheres.

Remark 2.6.10. It might be possible to push the techniques of this section to get more results on subgroups of hyperbolic groups. The following seem to be successively harder problems from this branched covers and Morse theory perspective.

1. Try to come up with a simpler (smaller number of cells) version of this example.

2. Try to take non-generic branched covers of complexes in higher dimensions to get hyperbolic groups with subgroups of type F_n but not F_{n+1} for $n \geqslant 3$.

3. Look for a hyperbolic group containing a subgroup which has a finite Eilenberg–Mac Lane space, but which is not hyperbolic.

K	$S(K)$	$T(K)$	X_K	A_K	H_K
•	• •		S^1	$\mathbb{Z}$	1
• •	• • • •		$S^1 \vee S^1$	F_2	$Ker(F_2 \to \mathbb{Z})$
1-simplex	$S^0 * S^0$		$S^1 \times S^1$	$\mathbb{Z}^2$	$\mathbb{Z}$

$S^0 * S^0$					
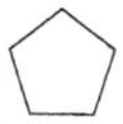					
2-simplex	$S^0 * S^0 * S^0$	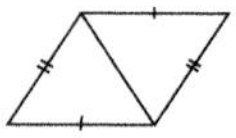	T^3	$\mathbb{Z}^3$	$\mathbb{Z}^2$
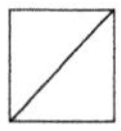					
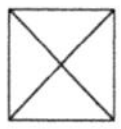					
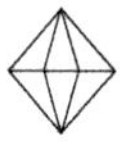					
$S^0 * S^0 * S^0$				$\displaystyle\prod_{i=1}^{3} F_2$	Stallings
3-simplex	$S^0 * \overset{4}{\cdots} * S^0$		T^4	$\mathbb{Z}^4$	$\mathbb{Z}^3$
n-simplex					
$S^0 * \cdots * S^0$					

Figure 2.15: Table of right-angled Artin group data. The exercise after Theorem 2.5.5 asks to complete this.

Bibliography

[1] J. Barnard and N. Brady. Distortion of surface groups in cat(0) free-by-cyclic groups. *Preprint*, 2006.

[2] G. Baumslag, M. Bridson, C. Miller, III, and H. Short. Finitely presented subgroups of automatic groups and their isoperimetric functions. *J. London Math. Soc. (2)* **56** (1997), no. 2, 292–304.

[3] G. Baumslag, C. Miller, III, and H. Short. Isoperimetric inequalities and the homology of groups. *Invent. Math.* **113** (1993), no. 3, 531–560.

[4] M. Bestvina and N. Brady. Morse theory and finiteness properties of groups. *Invent. Math.* **129** (1997), no. 3, 445–470.

[5] R. Bieri. Normal subgroups in duality groups and in groups of cohomological dimension 2. *J. Pure Appl. Algebra* **7** (1976), no. 1, 35–51.

[6] J. Birget, A. Ol'shanskii, E. Rips, and M. Sapir. Isoperimetric functions of groups and computational complexity of the word problem. *Ann. of Math. (2)* **156** (2002), no. 2, 476–518.

[7] B. Bowditch. A short proof that a subquadratic isoperimetric inequality implies a linear one. *Michigan Math. J.* 42 (1995), no. 1, 103–107.

[8] N. Brady. Branched coverings of cubical complexes and subgroups of hyperbolic groups. *J. London Math. Soc. (2)* **60** (1999), no. 2, 461–480.

[9] N. Brady and M. Bridson. There is only one gap in the isoperimetric spectrum. *Geom. Funct. Anal.* **10** (2000), no. 5, 1053–1070.

[10] N. Brady, M. Bridson, M. Forester, and K. Shankar. Perron–frobenius eigenvalues, snowflake groups, and isoperimetric spectra. *Preprint*, 2006.

[11] N. Brady, M. Forester, and K. Shankar. Dehn functions of subgroups of cat(0) groups. *Preprint*, 2006.

[12] N. Brady and A. Miller. CAT(−1) structures for free-by-free groups. *Geom. Dedicata* **90** (2002), 77–98.

[13] M. Bridson. Fractional isoperimetric inequalities and subgroup distortion. *J. Amer. Math. Soc.* **12** (1999), no. 4, 1103–1118.

[14] M. Bridson. Non-positive curvature in group theory. In *Groups St. Andrews 1997 in Bath, I*, volume 260 of *London Math. Soc. Lecture Note Ser.*, pp. 124–175. Cambridge Univ. Press, Cambridge, 1999.

[15] M. Bridson. The geometry of the word problem. In *Invitations to geometry and topology*, volume 7 of *Oxf. Grad. Texts Math.*, pp. 29–91. Oxford Univ. Press, Oxford, 2002.

[16] M. Bridson. Arkansas spring lecture series (lecture notes). 2006.

[17] M. Bridson and A. Haefliger. *Metric spaces of non-positive curvature*, volume 319 of *Grundlehren der Mathematischen Wissenschaften [Fundamental Principles of Mathematical Sciences]*. Springer-Verlag, Berlin, 1999.

[18] M. Bridson and C. Pittet. Isoperimetric inequalities for the fundamental groups of torus bundles over the circle. *Geom. Dedicata* **49** (1994), no. 2, 203–219.

[19] K. Brown. *Cohomology of groups*, volume 87 of *Graduate Texts in Mathematics*. Springer-Verlag, New York, 1982.

[20] K.-U. Bux and C. Gonzalez. The Bestvina-Brady construction revisited: geometric computation of Σ-invariants for right-angled Artin groups. *J. London Math. Soc. (2)* **60** (1999), no. 3, 793–801.

[21] M. Davis. Nonpositive curvature and reflection groups. In *Handbook of geometric topology*, pp. 373–422. North-Holland, Amsterdam, 2002.

[22] M. Elder and J. McCammond. Curvature testing in 3-dimensional metric polyhedral complexes. *Experiment. Math.* **11** (2002), no. 1, 143–158.

[23] M. Elder and J. McCammond. CAT(0) is an algorithmic property. *Geom. Dedicata* **107** (2004), 25–46.

[24] M. Elder, J. McCammond, and J. Meier. Combinatorial conditions that imply word-hyperbolicity for 3-manifolds. *Topology* **42** (2003), no. 6, 1241–1259.

[25] D. Epstein, J. Cannon, D. Holt, S. Levy, M. Paterson, and W. Thurston. *Word processing in groups*. Jones and Bartlett Publishers, Boston, MA, 1992.

[26] S. Gersten. Branched coverings of 2-complexes and diagrammatic reducibility. *Trans. Amer. Math. Soc.* **303** (1987), no. 2, 689–706.

[27] S. Gersten. Dehn functions and l_1-norms of finite presentations. In *Algorithms and classification in combinatorial group theory (Berkeley, CA, 1989)*, volume 23 of *Math. Sci. Res. Inst. Publ.*, pp. 195–224. Springer, New York, 1992.

[28] S. Gersten and H. Short. Some isoperimetric inequalities for kernels of free extensions. *Geom. Dedicata* **92** (2002), 63–72. Dedicated to John Stallings on the occasion of his 65th birthday.

[29] M. Gromov. Hyperbolic groups. In *Essays in group theory*, volume 8 of *Math. Sci. Res. Inst. Publ.*, pp. 75–263. Springer, New York, 1987.

[30] A. Hatcher. *Algebraic topology*. Cambridge University Press, Cambridge, 2002.

[31] J. Howie. On the asphericity of ribbon disc complements. *Trans. Amer. Math. Soc.* **289** (1985), no. 1, 281–302.

[32] A. Ol'shanskiĭ. Hyperbolicity of groups with subquadratic isoperimetric inequality. *Internat. J. Algebra Comput.* **1** (1991), no. 3, 281–289.

[33] P. Papasoglu. On the sub-quadratic isoperimetric inequality. In *Geometric group theory (Columbus, OH, 1992)*, volume 3 of *Ohio State Univ. Math. Res. Inst. Publ.*, pp. 149–157. de Gruyter, Berlin, 1995.

[34] M. Sapir, J.-C. Birget, and E. Rips. Isoperimetric and isodiametric functions of groups. *Ann. of Math. (2)* **156** (2002), no. 2, 345–466.

[35] P. Scott and T. Wall. Topological methods in group theory. In *Homological Group Theory (Proc. Sympos., Durham, 1977)*, London Math. Soc. Lecture Note Ser., pp. 137–203. Cambridge University Press, 1979.

[36] J. Stallings. A finitely presented group whose 3-dimensional integral homology is not finitely generated. *Amer. J. Math.* **85** (1963), 541–543.

[37] D. Wise. Cubulating small cancellation groups. *Geom. Funct. Anal.* **14** (2004), no. 1, 150–214.

Part II

Filling Functions

Tim Riley

Notation

$\preceq, \succeq, \simeq$	$f, g : [0, \infty) \to [0, \infty)$ satisfy $f \preceq g$ when there exists $C > 0$ such that $f(n) \le Cg(Cn+C)+Cn+C$ for all n, satisfy $f \succeq g$ when $g \preceq f$, and satisfy $f \simeq g$ when $f \preceq g$ and $g \preceq f$. These relations are extended to functions $f : \mathbb{N} \to \mathbb{N}$ by considering such f to be constant on the intervals $[n, n+1)$.
a^b, a^{-b}, $[a, b]$	$b^{-1}ab$, $b^{-1}a^{-1}b$, $a^{-1}b^{-1}ab$
$Cay^1(G, X)$	the Cayley graph of G with respect to a generating set X
$Cay^2(\mathcal{P})$	the Cayley 2-complex of a presentation $\mathcal{P}$
$\mathbb{D}^n$	the n-disc $\{(x_1, \ldots, x_n) \in \mathbb{R}^n \mid \sum_{i=1}^n x_i^2 \le 1\}$
$\mathrm{Diam}(\Gamma)$	$\max\{\rho(a, b) \mid$ vertices a, b in $\Gamma\}$, where ρ is the combinatorial metric on a finite connected graph Γ
d_X	the word metric with respect to a generating set X
$\ell(w)$	word length; i.e. the number of letters in the word w
$\ell(\partial\Delta)$	the length of the boundary circuit of Δ
$\mathbb{N}, \mathbb{R}, \mathbb{Z}$	the natural numbers, real numbers, and integers
R^{-1}	$\{r^{-1} \mid r \in R\}$, the inverses of the words in R
$\mathbb{S}^n$	the n-sphere, $\{(x_1, \ldots, x^{n+1}) \in \mathbb{R}^{n+1} \mid \sum_{i=1}^{n+1} = 1\}$
$\mathrm{Star}(\Delta_0)$	for a subcomplex $\Delta_0 \subseteq \Delta$, the union of all closed cells in Δ that have non-empty intersection with Δ_0
(T, T^*)	a dual pair of spanning trees — see Section 1.2
w^{-1}	the inverse $x_n^{-\varepsilon_n} \ldots x_2^{-\varepsilon_2} x_1^{-\varepsilon_1}$ of a word $w = x_1^{\varepsilon_1} x_2^{\varepsilon_2} \ldots x_n^{\varepsilon_n}$.
$\langle X \mid R \rangle$	the presentation with generators X and defining relations (or *relators*) R
X^{-1}	$\{x^{-1} \mid x \in X\}$, the formal inverses x^{-1} of letters x in an alphabet X
$(X \cup X^{-1})^*$	the free monoid (i.e. the words) on $X \cup X^{-1}$
Δ	a van Kampen diagram
ε	the empty word

Diagram measurements

$\mathrm{Area}(\Delta)$	the number of 2-cells in Δ
$\mathrm{DGL}(\Delta)$	$\min\{\mathrm{Diam}(T) + \mathrm{Diam}(T^*) \mid T$ a spanning tree in $\Delta^{(1)}\}$
$\mathrm{EDiam}(\Delta)$	the diameter of Δ as measured in the Cayley 2-complex
$\mathrm{FL}(\Delta)$	the filling length of Δ — see Section 1.2
$\mathrm{GL}(\Delta)$	the diameter of the 1-skeleton of the dual of Δ
$\mathrm{IDiam}(\Delta)$	the diameter of the 1-skeleton of Δ
$\mathrm{Rad}(\Delta)$	$\max\{\rho(a, \partial\Delta) \mid$ vertices a of $\Delta\}$ as measured in $\Delta^{(1)}$

Filling functions — see Section 1.2 unless otherwise indicated

$\mathrm{Area} : \mathbb{N} \to \mathbb{N}$	the Dehn function
$\mathrm{DGL} : \mathbb{N} \to \mathbb{N}$	the simultaneous diameter and gallery length function
$\mathrm{EDiam} : \mathbb{N} \to \mathbb{N}$	the extrinsic diameter function
$\mathrm{FL} : \mathbb{N} \to \mathbb{N}$	the filling length function
$\mathrm{GL} : \mathbb{N} \to \mathbb{N}$	the gallery length function
$\mathrm{IDiam} : \mathbb{N} \to \mathbb{N}$	the intrinsic diameter function
$\overline{\mathrm{IDiam}} : \mathbb{N} \to \mathbb{N}$	the upper intrinsic diameter function — see Section 2.1
$\mathrm{Rad} : \mathbb{N} \to \mathbb{N}$	the radius function — see Section 4.2
$\overline{\mathrm{Rad}} : \mathbb{N} \to \mathbb{N}$	the upper radius function — see Section 4.2

Introduction

The Word Problem was posed by Dehn [32] in 1912. He asked, given a group, for a systematic method (in modern terms, an *algorithm*) which, given a finite list (a *word*) of basic group elements (*generators* and their formal inverses), declares whether or not their product is the identity. One of the great achievements of 20th century mathematics was the construction by Boone [13] and Novikov [73] of finitely presentable groups for which no such algorithm can exist. However, the Word Problem transcends its origins in group theory and rises from defeat at the hands of decidability and complexity theory, to form a bridge to geometry — to the world of isoperimetry and curvature, local and large-scale invariants, as brought to light most vividly by Gromov [61].

So where does geometry enter? Groups act: given a group, one seeks a space on which it acts in as favourable a manner as possible, so that the group can be regarded as a discrete approximation to the space. Then a dialogue between geometry and algebra begins. And where can we find a reliable source of such spaces? Well, assume we have a finite generating set X for our group G. (All the groups in this study will be finitely generated.) For $x, y \in G$, define the distance $d_X(x, y)$ in the *word metric* d_X to be the length of the shortest word in the generators and their formal inverses that represents $x^{-1}y$ in G. Then

$$d_X(zx, zy) = d_X(x, y)$$

for all $x, y, z \in G$, and so left multiplication is action of G on (G, d_X) by isometries.

However (G, d_X) is discrete and so appears geometrically emaciated ("boring and uneventful to a geometer's eye" in the words of Gromov [61]). Inserting a directed edge labelled by a from x to y whenever $a \in X$ and $xa = y$ gives a skeletal structure known as the *Cayley graph* $Cay^1(G, X)$. If G is given by a finite presentation $\mathcal{P} = \langle X \mid R \rangle$ we can go further: attach flesh to the dry bones of $Cay^1(G, X)$ in the form of 2-cells, their boundary circuits glued along edge-loops around which read words in R. The result is a simply connected space $Cay^2(\mathcal{P})$ on which G acts *geometrically* (that is, properly, discontinuously and cocompactly) that is known as the *Cayley 2-complex* — see Section 1.1 for a precise definition. Further enhancements may be possible. For example, one could seek to attach cells of dimension 3 or above to kill off higher homotopy groups or so that the complex enjoys curvature conditions such as the CAT(0) property (see [21] or Part III of this volume).

Not content with the combinatorial world of complexes, one might seek smooth models for G. For example, one could realise G as the fundamental group of a closed manifold M, and then G would act geometrically on the universal cover $\widetilde{M}$. (If G is finitely presentable then M can be taken to be four dimensional — see [18, A.3].) Wilder non-discrete spaces, *asymptotic cones*, arise from viewing (G, d_X) from increasingly distant vantage points (*i.e.* scaling the metric to d_X/s_n for some sequence of reals with $s_n \to \infty$) and recording recurring patterns using the magic of a non-principal ultrafilter. Asymptotic cones discard all small-scale features of (G, d_X); they are the subject of Chapter 4.

Filling functions, the subject of this study, capture features of discs spanning loops in spaces. The best known is the classical *isoperimetric* function for Euclidean space $\mathbb{E}^m$ — any loop of length ℓ can be filled with a disc of area at most a constant times ℓ^2. To hint at how filling functions enter the world of discrete groups we mention a related algebraic result concerning the group $\mathbb{Z}^m$, the integer lattice in m-dimensional Euclidean space, generated by $x_1, \ldots, x_m$. If w is a word of length n on $\{x_1^{\pm 1}, \ldots, x_m^{\pm 1}\}$ and w represents the identity in $\mathbb{Z}^m$ then, by cancelling off pairs $x_i x_i^{-1}$ and $x_i^{-1} x_i$, and by interchanging adjacent letters at most n^2 times, w can be reduced to the empty word.

This qualitative agreement between the number of times the commutator relations $x_i x_j = x_j x_i$ are applied and the area of fillings is no coincidence; such a relationship holds for all finitely presented groups, as will be spelt out in Theorem 1.6.1 (*The Filling Theorem*). The bridge between continuous maps of discs filling loops in spaces and this computational analysis of reducing words is provided by *van Kampen* (or *Dehn*) *diagrams*. The Cayley 2-complex of the presentation

$$\mathcal{P} := \langle x_1, \ldots, x_m \mid [x_i, x_j], \forall i, j \in \{1, \ldots, m\} \rangle$$

of $\mathbb{Z}^m$ is the 2-skeleton of the standard decomposition of $\mathbb{E}^m$ into an infinite array of m-dimensional unit cubes. A word w that represents 1 in $\mathcal{P}$ (or, indeed, in any finite presentation $\mathcal{P}$) corresponds to an edge-loop in $Cay^2(\mathcal{P})$. As Cayley 2-complexes are simply connected such edge-loops can be spanned by filling discs and, in this combinatorial setting, it is possible and appropriate to take these homotopy discs to be combinatorial maps of planar 2-complexes homeomorphic to (possibly *singular*) 2-discs into $Cay^2(\mathcal{P})$. A *van Kampen diagram* for w is a graphical demonstration of how it is a consequence of the relations R that w represents 1; Figure 1.4 is an example. So the Word Problem amounts to determining whether or not a word admits a van Kampen diagram. (See *van Kampen's Lemma: Lemma 1.4.1.*)

Filling functions for finite presentations of groups (defined in Chapter 1) record geometric features of van Kampen diagrams. The best known is the *Dehn function* (or *minimal isoperimetric function*) of Madlener & Otto [68] and Gersten [46]; it concerns *area* — that is, number of 2-cells. In the example of $\mathbb{Z}^m$

this equates to the number of times commutator relations have to be applied to reduce w to the empty word — in this sense (that is, *in the Dehn proof system* — see Section 1.5) the Dehn function can also be understood as a non-deterministic TIME complexity measure of the Word Problem for $\mathcal{P}$ — see Section 1.5. The corresponding SPACE complexity measure is called the *filling length function* of Gromov [61]. It has a geometric interpretation — the filling length of a loop γ is the infimal length L such that γ can be contracted down to its base vertex through loops of length at most L. Other filling functions we will encounter include the *gallery length*, and *intrinsic* and *extrinsic diameter functions*. All are group invariants in that whilst they are defined with respect to specific finite presentations, their qualitative growth depends only on the underlying group; moreover, they are quasi-isometry invariants, that is, qualitatively they depend only on the large-scale geometry of the group — see Section 1.7 for details.

In Chapter 2 we examine the interplay between different filling functions — this topic bares some analogy with the relationships that exist between different algorithmic complexity measures and, as with that field, many open questions remain. The example of nilpotent groups discussed in Chapter 3 testifies to the value of simultaneously studying multiple filling functions. Finally, in Chapter 4, we discuss how the geometry and topology of the asymptotic cones of a group G relates to the filling functions of G.

Acknowledgements. These notes build on and complement [18], [47] and [61, Chapter 5] as well as the other two sets of notes in this volume. For Chapter 4 I am particularly indebted to the writings of Druţu [33, 34, 36, 37]. This is not intended to be a balanced or complete survey of the literature, but rather is a brief tour heavily biased towards areas in which the author has been involved.

If I have done any justice to this topic then the influence of Martin Bridson and Steve Gersten should shine through. I am grateful to them both for stimulating collaborations, for their encouragement, and for communicating their deep vision for the subject. I thank Emina Alibegovic, Will Dison, Cornelia Druţu Badea, Steve Gersten, Mark Sapir and Hamish Short for comments on earlier drafts. I also thank the NSF for partial support via grant DMS–0540830.

Tim Riley

Chapter 1

Filling Functions

1.1 Van Kampen diagrams

The *presentation 2-complex* of

$$\mathcal{P} = \langle X \mid R \rangle = \langle x_1, \ldots, x_m \mid r_1, \ldots, r_n \rangle$$

is constructed as shown in Figure 1.1: take m oriented edges, labelled by $x_1, \ldots, x_m$, identify all the vertices to form a *rose*, and then attach 2-cells $C_1, \ldots, C_n$, where C_i has $\ell(r_i)$ edges, by identifying the boundary circuit of C_i with the edge-path in the rose along which one reads r_i. The *Cayley 2-complex* $Cay^2(\mathcal{P})$ is the universal cover of the presentation 2-complex of $\mathcal{P}$. The example of the free abelian group of rank 2, presented by $\langle a, b \mid [a, b] \rangle$ is shown in Figure 1.2.

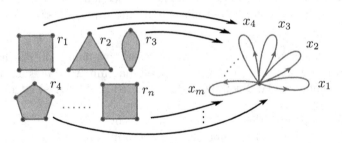

Figure 1.1: The presentation 2-complex for $\langle x_1, \ldots, x_m \mid r_1, \ldots, r_n \rangle$.

The edges of $Cay^2(\mathcal{P})$ inherit labels and orientations from the presentation 2-complex and the 1-skeleton of $Cay^2(\mathcal{P})$ is the Cayley graph $Cay^1(\mathcal{P})$ (cf. Definition 1.3 in Part I of this volume). Identifying the group G presented by $\mathcal{P}$ with the 0-skeleton of $Cay^1(\mathcal{P})$, the path metric on $Cay^1(\mathcal{P})$ in which each edge has length 1 agrees with the word metric d_X on G.

A word $w \in (X \cup X^{-1})^*$ is *null-homotopic* when it represents the identity. To such a w one can associate an edge-circuit η_w in $Cay^2(\mathcal{P})$ based at some (and

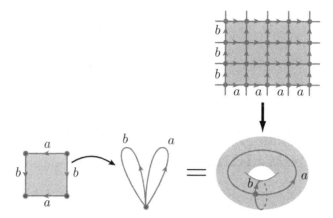

Figure 1.2: The presentation and Cayley 2-complexes for $\langle a, b \mid [a, b] \rangle$.

hence any) fixed vertex v, so that around η_w, starting from v, one reads w. A *$\mathcal{P}$-van Kampen diagram* for w is a combinatorial map $\pi : \Delta \to Cay^2(\mathcal{P})$, where Δ is $S^2 \smallsetminus e_\infty$ for some combinatorial 2-complex S^2 homeomorphic to the 2-sphere and some open 2-cell e_∞ of S^2, and the edge-circuit $\pi \mid_{\partial\Delta}$, based at vertex $\star$ on $\partial\Delta$, agrees with η_w. (A map $\mathcal{X} \to \mathcal{Y}$ between complexes is *combinatorial* if for all n, it maps the interior of each n-cell in $\mathcal{X}$ homeomorphically onto the interior of an n-cell in $\mathcal{Y}$.) The base vertex $\star$ of Δ should not be ignored; we will see it plays a crucial role. Van Kampen's Lemma (see Section 1.4) says, in particular, that $w \in (X \cup X^{-1})^*$ is null-homotopic if and only if it admits a $\mathcal{P}$-van Kampen diagram.

It is convenient to have the following alternative definition in which no explicit reference is made to π. A *$\mathcal{P}$-van Kampen diagram* Δ for w is a finite planar contractible combinatorial 2-complex with directed and labelled edges such that anti-clockwise around $\partial\Delta$ one reads w, and around the boundary of each 2-cell one reads a cyclic conjugate (that is, a word obtained by cyclically permuting letters) of a word in $R^{\pm 1}$. From this point of view a $\mathcal{P}$-van Kampen diagram for w is a filling of a planar edge-loop labelled by w in the manner of a jigsaw-puzzle, where the pieces (such as those pictured in Figure 1.3) correspond to defining relations and are required to be fitted together in such a way that orientations and labels match. The analogy breaks down in that the pieces may be distorted or flipped. Figures 1.4, 1.7, 2.8 and 3.1 show examples of van Kampen diagrams.

The two definitions are, in effect, equivalent because, given the first, edges of Δ inherit directions and labels from $Cay^2(\mathcal{P})$, and given the second, there is a combinatorial map $\pi : \Delta \to Cay^2(\mathcal{P})$, uniquely determined on the 1-skeleton of Δ, that sends $\star$ to v and preserves the labels and directions of edges.

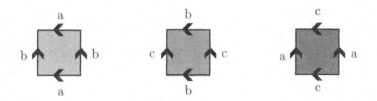

Figure 1.3: The defining relations in $\langle a, b, c \mid [a, b], [b, c], [c, a] \rangle$.

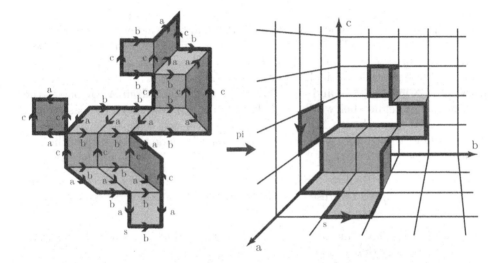

Figure 1.4: A van Kampen diagram for
$ba^{-1}ca^{-1}bcb^{-1}ca^{-1}b^{-1}c^{-1}bc^{-1}b^{-2}acac^{-1}a^{-1}c^{-1}aba$ in $\langle a, b, c \mid [a, b], [b, c], [c, a] \rangle$.

1.2 Filling functions via van Kampen diagrams

Suppose $\mathcal{P} = \langle X \mid R \rangle$ is a finite presentation for a group and $\pi : \Delta \to Cay^2(\mathcal{P})$ is a van Kampen diagram. Being planar, Δ has a dual Δ^* — the 2-complex with a vertex, edge and face dual to each face, edge and vertex in Δ (including a vertex dual to the face *at infinity*, the complement of Δ in the plane).

For a finite connected graph Γ define

$$\mathrm{Diam}(\Gamma) := \max \{ \, \rho(a, b) \mid \text{vertices } a, b \text{ of } \Gamma \, \}$$

where ρ is the combinatorial metric on Γ — the path metric in which each edge is given length 1. Define the *area, intrinsic diameter, extrinsic diameter, gallery length, filling length* and DGL of $\pi : \Delta \to Cay^2(\mathcal{P})$ by

$$
\begin{aligned}
\text{Area}(\Delta) &= \#\ 2\text{-cells in }\Delta \\
\text{IDiam}(\Delta) &= \text{Diam}(\Delta^{(1)}) \\
\text{EDiam}(\Delta) &= \max\{\ d_X(\pi(a),\pi(b))\ \mid\ \text{vertices }a,b\text{ of }\Delta\ \} \\
\text{GL}(\Delta) &= \text{Diam}(\Delta^{\star(1)}) \\
\text{FL}(\Delta) &= \min\left\{\ L\ \middle|\ \exists\text{ a }\textit{shelling}\ (\Delta_i)\text{ of }\Delta\text{ such that }\max_i \ell(\partial\Delta_i) \le L\ \right\} \\
\text{DGL}(\Delta) &= \min\left\{\ \text{Diam}(T) + \text{Diam}(T^*)\ \middle|\ T\text{ a spanning tree in }\Delta^{(1)}\ \right\}.
\end{aligned}
$$

These are collectively referred to as *diagram measurements*. Note that IDiam measures diameter in the 1-skeleton of Δ and EDiam in the Cayley graph of $\mathcal{P}$.

The definitions of $\text{FL}(\Delta)$ and $\text{DGL}(\Delta)$ require further explanation. A *shelling* of Δ is, roughly speaking, a *combinatorial null-homotopy* of Δ down to its base vertex $\star$. More precisely, a shelling of Δ is sequence $\mathcal{S} = (\Delta_i)$ of diagrams $(\Delta_i)_{i=0}^{m}$ with $\Delta_0 = \Delta$ and $\Delta_m = \star$ and such that Δ_{i+1} is obtained from Δ_i by one of the *shelling moves* defined below (illustrated in Figure 1.5). Moreover, the base vertex $\star$ is *preserved* throughout (Δ_i) — that is, in every 1-*cell collapse* $e^0 \ne \star$, and in every 1-*cell expansion* on Δ_i where $e_0 = \star$ a choice is made as to which of the two copies of e_0 is to be $\star$ in Δ_{i+1}.

- 1-*cell collapse.* Remove a pair (e^1, e^0) where e^1 is a 1-cell with $e^0 \in \partial e^1$ and e^1 is attached to the rest of Δ_i only by one of its end vertices $\ne e^0$. (Call such an e^1 a *spike.*)

- 1-*cell expansion.* Cut along some 1-cell e^1 in Δ_i that has a vertex e^0 in $\partial\Delta_i$, in such a way that e^0 and e^1 are doubled.

- 2-*cell collapse.* Remove a pair (e^2, e^1) where e^2 is a 2-cell which has some edge $e^1 \in (\partial e^2 \cap \partial\Delta_i)$. The effect on the boundary circuit is to replace e^1 with $\partial e^2 \smallsetminus e^1$.

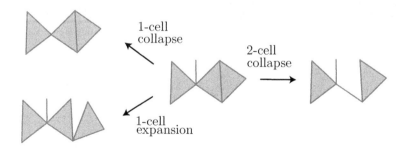

Figure 1.5: Shelling moves

There are natural combinatorial maps $\Delta_i \to \Delta$ whose restrictions to the interiors of Δ_i are injective, and which map the boundary circuits of Δ_i, labelled by words w_i, to a sequence of edge-loops contracting to $\star$ as illustrated schematically in Figure 1.6.

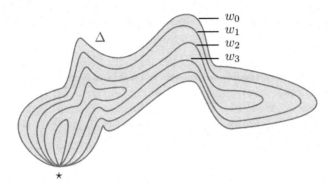

Figure 1.6: The contracting loops in the course of a null-homotopy

Returning to the definition of $\mathrm{DGL}(\Delta)$, if T is a spanning tree in the 1-skeleton of Δ, then define T^* to be the subgraph of the 1-skeleton of Δ^* consisting of all the edges dual to edges of $\Delta^{(1)} \smallsetminus T$. The crucial property of T^* is:

Exercise 1.2.1. T^* is a spanning tree in the 1-skeleton of Δ^*.

We now define an assortment of *filling functions* for $\mathcal{P}$.

- The *Dehn function* $\mathrm{Area} : \mathbb{N} \to \mathbb{N}$,

- the *intrinsic diametric function* $\mathrm{IDiam} : \mathbb{N} \to \mathbb{N}$,

- the *extrinsic diametric function* $\mathrm{EDiam} : \mathbb{N} \to \mathbb{N}$,

- the *gallery length function* $\mathrm{GL} : \mathbb{N} \to \mathbb{N}$,

- the function $\mathrm{DGL} : \mathbb{N} \to \mathbb{N}$,

- and the *filling length function* $\mathrm{FL} : \mathbb{N} \to \mathbb{N}$

of $\mathcal{P}$ are defined by

$$\mathrm{M}(w) \; := \; \min\{\, \mathrm{M}(\Delta) \; | \; \Delta \text{ a van Kampen diagram for } w \,\},$$
$$\mathrm{M}(n) \; := \; \max\{\, \mathrm{M}(w) \; | \; \text{words } w \text{ with } \ell(w) \leq n \text{ and } w = 1 \text{ in } G \,\},$$

where M is Area, IDiam, EDiam, GL, DGL and FL respectively. (The meanings of $\mathrm{M}(\Delta)$, $\mathrm{M}(w)$ and $\mathrm{M}(n)$ depend on their arguments: diagram, null-homotopic word, or natural number; the potential ambiguity is tolerated as it spares us from an overload of terminology.)

An *isoperimetric function* (respectively, *isodiametric function*) for $\mathcal{P}$ is any $f : \mathbb{N} \to \mathbb{N}$ such that $\mathrm{Area}(n) \leq f(n)$ (respectively, $\mathrm{IDiam}(n) \leq f(n)$) for all n. Chapter 5 of [61] is a foundational reference on filling functions. Many other references to isoperimetric functions, Dehn functions and isodiametric functions appear in the literature; [18] and [47] are surveys. Dehn functions were introduced by Madlener & Otto [68] and, independently, by Gersten [46]. The filling length function of Gromov [61] is discussed extensively in [52]; a closely related notion LNCH was introduced by Gersten [48]. Gallery length and DGL were introduced in [51], and EDiam appears in [22].

Definition 1.2.2. We say that functions $f_i : \mathbb{N} \to \mathbb{N}$ are *simultaneously realisable* (upper or lower) bounds on a collection of filling functions $M_i : \mathbb{N} \to \mathbb{N}$ of $\mathcal{P}$ (that is, each M_i is one of Area, FL, IDiam,...) if for every null-homotopic word w, there exists a van Kampen diagram Δ for w such that $f_i(\ell(w))$ is at most or at least (as appropriate) $M_i(\Delta)$, for all i.

The following two exercises are essentially elementary observations. The second serves to describe a combinatorial group theoretic adaptation from [54] of a variant F_+L of filling length defined by Gromov in [61, page 101]. F_+L was used in [54] to show that groups that enjoy Cannon's *almost convexity condition* AC(2) have filling length functions (in the standard sense) growing $\preceq n$.

Exercise 1.2.3. Show that for a finite presentation, $d(u,v) := \mathrm{Area}(u^{-1}v)$ defines a metric on any set of reduced words, all representing the same group element.

Exercise 1.2.4. Suppose $u, v \in (X^{\pm 1})^*$ represent the same group element and Δ is a van Kampen diagram for uv^{-1} with two distinguished boundary vertices $\star_1, \star_2$ separating the u- and v-portions of the boundary circuit. Define $F_+L(u, v, \Delta)$ to be the least L such that there is a *combinatorial homotopy of u to v across* Δ through paths of length at most L from $\star_1$ to $\star_2$. (More formally, such a combinatorial homotopy is a sequence of van Kampen diagrams $(\Delta_i)_{i=0}^m$ with $\Delta_0 = \Delta$ and Δ_m a simple edge-path along which one reads w_2, and such that Δ_{i+1} is obtained from Δ_i by 1-*cell collapse*, 1-*cell expansion*, or 2-*cell collapse* in such a way that the v-portion of the boundary words $\partial \Delta_i$ is not broken.) Define $F_+L(u, v)$ to be the minimum of $F_+L(u, v, \Delta)$ over all van Kampen diagrams Δ for uv^{-1} and define $F_+L : \mathbb{N} \to \mathbb{N}$ by letting $F_+L(n)$ be the maximum of $F_+L(u, v)$ over all u, v of length at most n that represent the same element of the group. Show that

1. $FL(2n) \le F_+L(n)$ and $FL(2n + 1) \le F_+L(n + 1)$ for all n.

2. $F_+L(u, v)$ defines a metric on any set of words that all represent the same group element.

1.3 Example: combable groups

Combable groups form a large and well-studied class that includes all automatic groups [41] and (hence) all hyperbolic groups. A *normal form* for a group G with finite generating set X is a section $\sigma : G \to (X \cup X^{-1})^*$ of the natural surjection $(X \cup X^{-1})^* \twoheadrightarrow G$. In other words, a normal form is a choice of representative $\sigma_g = \sigma(g)$ for each group element g. View σ_g as a continuous path $[0, \infty) \to Cay^1(G, X)$

from the identity to g (the "*combing line* of g"), travelling at unit speed from the identity until time $\ell(\sigma_g)$ when it halts for evermore at g.

Following [15, 47], we say σ *synchronously k-fellow-travels* when

$$\forall g, h \in G, \quad (d_X(g, h) = 1 \implies \forall t \in \mathbb{N}, \ d_X(\sigma_g(t), \sigma_h(t)) \leq k).$$

Define a *reparametrisation* ρ to be an unbounded function $\mathbb{N} \to \mathbb{N}$ such that $\rho(0) = 0$ and $\rho(n+1) \in \{\rho(n), \rho(n) + 1\}$ for all n. We say σ *asynchronously k-fellow travels* when for all $g, h \in G$ with $d_X(g, h) = 1$, there exist reparametrisations ρ and ρ' such that

$$\forall t \in \mathbb{N}, \quad d_X(\sigma_g(\rho(t)), \sigma_h(\rho'(t))) \leq k.$$

(Note ρ and ρ' both depend on both g and h.)

We say (G, X) is *(a)synchronously combable* when, for some $k \geq 0$, there is an (a)synchronous k-fellow-travelling normal form σ for G. Define the *length function* $\mathrm{L} : \mathbb{N} \to \mathbb{N}$ of σ by:

$$\mathrm{L}(n) := \max \{ \ell(\sigma_g) \mid d_X(1, g) \leq n \}.$$

Examples 1.3.1.

1. *Finite groups.* Suppose G is a finite group with finite generating set X. Let k be the diameter of $Cay^1(G, X)$. If for all $g \in G$ we take σ_g to be any word representing g, then σ is synchronously k-fellow travelling. Moreover, if σ_g is a choice of geodesic (i.e. minimal length) word representing g, then σ has length function satisfying $\mathrm{L}(n) \leq \min \{n, k\}$ for all n.

2. The (unique) geodesic words representing elements in $F_m = \langle a_1, \ldots, a_m \mid \rangle$ form a synchronous 1-fellow-travelling combing. The words

$$\{ a_1^{r_1} a_2^{r_2} \ldots a_m^{r_m} \mid r_1, \ldots, r_m \in \mathbb{Z} \}$$

comprise a synchronous 2-fellow travelling combing of

$$\mathbb{Z}^m = \langle a_1, \ldots, a_m \mid [a_i, a_j], \forall 1 \leq i < j \leq m \rangle.$$

In both cases $L(n) = n$ for all n.

3. The family of groups $\mathrm{BS}(m, n)$ with presentations $\langle a, b \mid b^{-1} a^m b = a^n \rangle$ are named in honour of Baumslag and Solitar who studied them in [7]. $\mathrm{BS}(1, 2)$ is often referred to as "the Baumslag–Solitar group" despite repeated public insistence from Baumslag that this is an inappropriate attribution.

   The words

   $$\left\{ b^r u a^s \mid r, s \in \mathbb{Z}, \ u \in \{ab^{-1}, b^{-1}\}^*, \text{ the first letter of } u \text{ is not } b^{-1} \right\}$$

   define an asynchronous combing of $\mathrm{BS}(1, 2)$. In fact, the normal form element for $g \in \mathrm{BS}(1, 2)$ results from applying the *rewriting rules*

$$
\begin{array}{rcl}
ab & \mapsto & ba^2 \\
a^{-1}b & \mapsto & ba^{-2} \\
a^2b^{-1} & \mapsto & b^{-1}a \\
a^{-1}b^{-1} & \mapsto & ab^{-1}a^{-1}
\end{array}
\qquad\qquad
\begin{array}{rcl}
aa^{-1} & \mapsto & \varepsilon \\
a^{-1}a & \mapsto & \varepsilon \\
bb^{-1} & \mapsto & \varepsilon \\
b^{-1}b & \mapsto & \varepsilon
\end{array}
$$

to any word representing g. See [41, Chapter 7] for more details.

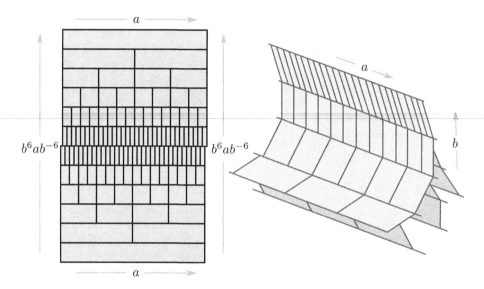

Figure 1.7: A van Kampen diagram for $\left[a, a^{b^6}\right]$ in $\langle a, b \mid b^{-1}ab = a^2 \rangle$ and a portion of $Cay^2(\mathcal{P})$.

4. *Automatic groups.* A group G with finite generating set X is *synchronously automatic* when it admits a synchronous combing σ such that the set of normal forms $\{\sigma_g \mid g \in G\}$ comprise a *regular language*. The foundational reference is [41] — there, a group is defined to be synchronous automatic when its multiplication and inverse operations are governed by finite state automata, and agreement of the two definitions is Theorem 2.3.5.

The σ_g are quasi-geodesics and so there is a constant C such that $\mathrm{L}(n) \leq Cn + C$ for all n. Synchronously automatic groups include all finitely generated virtually abelian groups, all hyperbolic groups, all the braid groups B_n, many 3-manifold groups, and the mapping class groups of closed surfaces with finitely many punctures — see [41, 72].

The larger class of *asynchronously automatic groups*, defined in [41], can be characterised (see Theorems 7.2.8 and 7.3.6 in [41]) as the asynchronously combable groups G such that $\{\sigma_g \mid g \in G\}$ is regular and admits a function $D : \mathbb{N} \to \mathbb{N}$ such that for all $g \in G$, $r, s \geq 0$, $t \geq D(r)$, if $s + t \leq \ell(\sigma_g)$

then $d_X(\sigma_g(s), \sigma_g(s+t)) > r$. Their length functions satisfy $L(n) \leq C^n$ for some $C > 0$ (see e.g. [19, Lemma 2.3]). Thurston [41, Section 7.4] showed that $BS(1,2)$ is asynchronously automatic: the language of the asynchronous combing given above is regular but, on account of of the exponential Dehn function we will establish in Proposition 2.1.6, $BS(1,2)$ is not synchronously automatic.

For references of an introductory nature and for open questions see, e.g. [6, 42, 48, 49, 74].

5. CAT(0) *groups.* Paths in a Cayley graph of a CAT(0) group that run close to CAT(0) geodesics can be used to define a synchronous combing with length function $L(n) \simeq n$. See the proof of Proposition 1.6 in [21, III.Γ] for more details.

Further classes of groups can be defined by specifying other grammatical, length function or geometric constraints. See, for example, [20, 42, 57] for more details.

Exercise 1.3.2. Check that the words in Example 1.3.1(3) define an asynchronous combing of $BS(1,2)$ and show that its length function satisfies $L(n) \simeq 2^n$. Estimate the fellow-travelling constant.

Exercise 1.3.3. Show that if X and X' are two finite generating sets for a group G then (G, X) is (a)synchronously combable if and only if (G, X') is. Moreover, show that if $L(n)$ is a length function for an (a)synchronous combing of (G, X) then there is an (a)synchronous combing of (G, X') with length function $L'(n)$ satisfying $L(n) \simeq L'(n)$.

Theorem 1.3.4 (cf. [41, 47, 51]). *Suppose a group G with finite generating set X admits a combing σ that (a)synchronously k-fellow-travels. Let $L : \mathbb{N} \to \mathbb{N}$ be the length function of σ. Then there exists $C > 0$ and a finite presentation $\mathcal{P} = \langle X \mid R \rangle$ for G for which*

$$\text{Area}(n) \;\leq\; C n \, L(n) \quad and$$
$$\text{EDiam}(n) \;\leq\; \text{IDiam}(n) \;\leq\; \text{FL}(n) \;\leq\; C n$$

for all n. Moreover, these bounds are realisable simultaneously.

Proof. Suppose w is a word representing 1 in G. We will construct the *cockleshell* van Kampen diagram Δ for w, illustrated in Figure 1.8. We start with a circle C in the plane subdivided into $\ell(w)$ edges, directed and labelled so that anticlockwise from the base vertex $\star$ one reads w. Then we join each vertex v to $\star$ by an edge-path labelled by the word σ_{g_v}, where g_v is the group element represented by the prefix of the word w read around C between $\star$ and v. For each pair u, v of adjacent vertices on C, we use the fact that σ_{g_u} and σ_{g_v} asynchronously k-fellow travel (with respect to reparametrisations ρ and ρ') to construct a *ladder* whose *rungs* are paths of length at most k; these would be straight if our combing was synchronous, but may be skew in the asynchronous case. A rung connects $\sigma_{g_u}(\rho(t))$ to $\sigma_{g_v}(\rho'(t))$ for $t = 1, 2, \ldots$. Duplicate rungs may occur on account of ρ and ρ' simultaneously pausing, that is, $\rho(m) = \rho(m+1)$ and $\rho'(m) = \rho'(m+1)$ for some m; counting each of these only once, there are a total of at most $\ell_{g_u} + \ell_{g_v}$ rungs and hence at most the same number of 2-cells in the ladder.

Subdivide the rung from $\sigma_{g_u}(\rho(t))$ to $\sigma_{g_v}(\rho'(t))$ into $d_X(\sigma_{g_u}(\rho(t)), \sigma_{g_v}(\rho'(t)))$ $\leq k$ edges and label it by a shortest word representing the same element in G as $[\sigma_{g_u}(\rho(t))]^{-1}\sigma_{g_v}(\rho'(t))$.

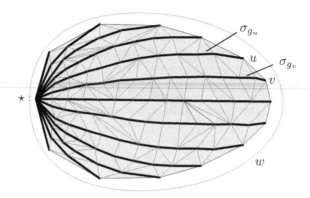

Figure 1.8: The cockleshell diagram filling an edge-loop in an asynchronously combable group.

Obtain Δ by regarding every simple edge-circuit in a ladder to be the boundary circuit of a 2-cell; words around such edge-circuits have length at most $2k+2$ and represent the identity in G. So, defining R to be the set of null-homotopic words of length at most $2k+2$, we find that $\mathcal{P} := \langle X \mid R \rangle$ is a finite presentation for G and Δ is a $\mathcal{P}$-van Kampen diagram for w. We get the asserted bound on Area(n) from the observation

$$\mathrm{Area}(\Delta) \ \leq \ 2\,\ell(w)\,\mathrm{L}(\lfloor\ell(w)/2\rfloor).$$

We will only sketch a proof of the linear upper bound on FL(n). Consider paths p connecting pairs of points on $\partial\Delta$, that run along successive rungs of adjacent ladders, no two rungs coming from the same ladder. Observe that there is such a p for which the region enclosed by p and a portion of the boundary circuit of Δ not passing through $\star$ is a topological 2-disc that includes no rungs in its interior. Shell away this region using 1-cell and 2-cell collapse moves. Repeating this process and performing 1-cells collapse moves will shell Δ down to $\star$ and the boundary loops of the intermediate diagrams will run (roughly speaking) perpendicular to the combing lines.

The inequalities EDiam(n) $\leq$ IDiam(n) $\leq$ FL(n) are straightforward and are not specific to combable groups — see Chapter 2. $\square$

Remark 1.3.5. Figure 1.8 suggests that $\text{DGL}(\Delta)$ and (hence) $\text{GL}(\Delta)$ are both at most a constant C times $\text{L}(\lfloor \ell(w)/2 \rfloor)$, taking T (a spanning tree in the 1-skeleton of Δ) to be the union of all the combing lines, and hence that there is a further simultaneously realisable bound of

$$\text{GL}(n) \leq \text{DGL}(n) \leq C\text{L}(n),$$ (1.1)

for some constant C. But the picture is misleading — some of the rungs could have zero length and hence T could fail to be a tree. A way to circumvent this problem is to *fatten* $\mathcal{P}$, that is, introduce a spurious extra generator z, that will be trivial in the group and add, in a sense redundant, defining relations $z, z^2, z^3, zz^{-1}, z^2z^{-1}$ and $[z, x]$ for all $x \in X$. The combing lines can be kept apart in this new presentation and (1.1) then holds. Details are in [51].

Remark 1.3.6. It is tempting to think that an induction on word length would show that the length function of a k-fellow-travelling synchronous combing satisfies $L(n) \leq Cn$ for some constant $C = C(k)$. But this is not so: Bridson [20] gave the first example of a synchronously combable group with Dehn function growing faster than quadratic — his example has cubic Dehn function. In particular, on account of this Dehn function, his group is not automatic.

1.4 Filling functions interpreted algebraically

By definition, when a group G is presented by $\langle X \mid R \rangle$ there is a short exact sequence, $\langle\langle R \rangle\rangle \hookrightarrow F(X) \twoheadrightarrow G$, where $\langle\langle R \rangle\rangle$ is the normal closure of R in $F(X)$. So a word $w \in (X \cup X^{-1})^*$ represents 1 in G if and only if there is an equality

$$w = \prod_{i=1}^{N} u_i^{-1} r_i^{\epsilon_i} u_i$$ (1.2)

in $F(X)$ for some integer N, some $\epsilon_i = \pm 1$, some words u_i, and some $r_i \in R$. Recent surveys including proofs of the following key lemma are [18] (Theorem 4.2.2) and Part I of this volume (Theorem 2.2).

Lemma 1.4.1 (Van Kampen's Lemma [101]). *A word w is null-homotopic in a finite presentation $\mathcal{P}$ if and only if it admits a $\mathcal{P}$-van Kampen diagram. Moreover, Area(w) is the minimal N such that there is an equality in $F(X)$ of the form (1.2).*

Exercise 1.4.2. Adapt a proof of van Kampen's Lemma to show that

$$L \leq \text{IDiam}(w) \leq 2L + \max\{\ell(r) \mid r \in R\} + \ell(w),$$

where L is the minimal value of $\max_i \ell(u_i)$ amongst all equalities (1.2) in $F(X)$. (The additive term $\ell(w)$ can be discarded when w is freely reduced.)

Exercise 1.4.3. Similarly, relate EDiam(w) to the minimal value of

$$\max \left\{ d_X(1, w_0) \;\middle|\; \text{prefixes } w_0 \text{ of the unreduced word } \prod_{i=1}^{N} u_i^{-1} r_i^{\epsilon_i} u_i \right\}$$

amongst all equalities (1.2) in $F(X)$.

1.5 Filling functions interpreted computationally

Given a finite presentation $\mathcal{P} = \langle X \mid R \rangle$ and a word $w_0 \in (X \cup X^{-1})^*$ imagine writing w_0 on the tape of a Turing machine, one letter in each of $\ell(w_0)$ adjacent squares. The moves described in the following lemma are ways of altering the word on the tape. Using these moves with the aim of making all squares blank is the *Dehn proof system* for trying to show w represents the identity in $\mathcal{P}$. (The author recalls hearing this terminology in a talk by Razborov, but is unaware of its origins.) We call $(w_i)_{i=0}^m$ a *null-$\mathcal{P}$-sequence* for w_0 (or a *$\mathcal{P}$-sequence* if we remove the requirement that the final word w_m be the empty word).

Lemma 1.5.1. *Suppose $w_0 \in (X \cup X^{-1})^*$. Then w_0 represents 1 in $\mathcal{P} = \langle X \mid R \rangle$ if and only if it can be reduced to the empty word w_m via a sequence $(w_i)_{i=0}^m$ in which each w_{i+1} is obtained from w_i by one of three moves:*

1. Free reduction: $w_i = \alpha x x^{-1} \beta \mapsto \alpha\beta = w_{i+1}$, *where $x \in X^{\pm 1}$.*

2. Free expansion: *the inverse of a free reduction.*

3. Application of a relator: $w_i = \alpha u \beta \mapsto \alpha v \beta = w_{i+1}$, *where a cyclic conjugate of uv^{-1} is in $R^{\pm 1}$.*

Proof. If w_0 is null-homotopic then, by Lemma 1.4.1, it admits a van Kampen diagram Δ_0. Let $(\Delta_i)_{i=0}^m$ be any shelling of Δ_0 down to its base vertex $\star$ by 1-cell- and 2-cell-collapse moves. Let w_i be the boundary word of Δ_i read from $\star$. Then $(w_i)_{i=0}^m$ is a null-sequence for w_0. The converse is straight-forward.   □

Proposition 1.5.2. *Suppose w_0 is null-homotopic in $\mathcal{P}$. Then, amongst all null-sequences (w_i) for w_0, Area(w_0) is the minimum A of the number of i such that w_{i+1} is obtained from w_i by an* application-of-a-relator *move, and FL(w_0) is the minimum of $\max_i \ell(w_i)$.*

Sketch proof. The proof of Lemma 1.5.1 shows $A \leq$ Area(w_0) and $F \leq$ FL(w_0).

For the reverse inequalities suppose (w_i) is a null-sequence for w_0 and that A and F are the number of i such that w_{i+1} is obtained from w_i by an *application-of-a-relator* move, and of $\max_i \ell(w_i)$, respectively. We seek a van Kampen diagram Δ_0 for w_0 such that Area(w_0) $\leq A$ and FL(w_0) $\leq F$. We will describe a sequence of planar diagrams A_i that topologically are singular annuli and have the property that the outer boundary of A_i is labelled by w_0 and the inner boundary by w_i.

Start with a planar, simple edge-circuit A_0 labelled by w_0. Obtain A_{i+1} from A_i as follows. If $w_i \mapsto w_{i+1}$ is a free reduction $w_i = \alpha x x^{-1} \beta \mapsto \alpha \beta = w_{i+1}$ then identify the two appropriate adjacent edges labelled x and x^{-1} in A_i. If it is a free expansion $w_i = \alpha \beta \mapsto \alpha x x^{-1} \beta = w_{i+1}$, then attach an edge labelled by x by its initial vertex to the inner boundary circuit of A_i. If it is an application of a relator $w_i = \alpha u \beta \mapsto \alpha v \beta = w_{i+1}$ then attach a 2-cell with boundary labelled by $u^{-1} v$ along the u portion of the inner boundary circuit of A_i.

However, the problem with this method of obtaining A_{i+1} from A_i is that there is no guarantee that A_{i+1} will be planar. For example, if $w_i \mapsto w_{i+1}$ is a free expansion immediately reversed by a free reduction $w_{i+1} \mapsto w_{i+2}$, then an edge will be pinched off. The resolution is that moves that give rise to such problems are redundant and (w_i) can be amended to avoid the difficulties, and this can be done without increasing A or F. A careful treatment is in [52]. □

The proof outlined above actually shows more: if (w_i) is a null-sequence for w then there is a van Kampen diagram Δ for w that simultaneously has area at most the number of application-of-a-relator moves in (w_i) and has $\mathrm{FL}(\Delta) \leq \max_i \ell(w_i)$.

Examples 1.5.3.

1. *Hyperbolic groups.* A group G with finite generating set X is (*Gromov-*) *hyperbolic* if and only if $Cay^1(G, X)$ is hyperbolic in the sense we will define in Section 4.2. This condition turns out to be independent of the finite generating set X (as follows from Theorem 4.2.2, for example). Equivalently [21, page 450], a group is hyperbolic if and only if it admits a *Dehn presentation* — a finite presentation $\mathcal{P}$ such that if w is null-homotopic, then one can perform an application-of-a-relator move $w = \alpha u \beta \mapsto \alpha v \beta$ for which $\ell(v) < \ell(u)$. It follows that $\mathrm{Area}(n), \mathrm{FL}(n) \leq n$, and these bounds are simultaneously realisable.

   So the Dehn and filling length functions of hyperbolic groups grow at most linearly.

2. *Finitely generated abelian groups.* Suppose G is a finitely generated abelian group
$$\langle\, x_1, \ldots, x_m \mid x_i^{\alpha_i}, [x_i, x_j], \forall 1 \leq i < j \leq m \,\rangle$$
   where each $\alpha_i \in \{0, 1, 2, \ldots\}$. Any word representing the identity can be reduced to the empty word by shuffling letters past each other to collect up powers of x_i, for all i, and then freely reducing and applying the $x_i^{\alpha_i}$ relators. So $\mathrm{Area}(n) \preceq n^2$ and $\mathrm{FL}(n) \leq n$ and these bounds are simultaneously realisable.

Implementing the Dehn Proof System on a Turing machine shows the following. [A Turing machine is *symmetric* when the transition relation is symmetric: the reverse of any transition is again a transition.]

Theorem 1.5.4 ([95, Theorem 1.1], [11, Proposition 5.2]). *For every finite presentation $\mathcal{P}$, there is a symmetric Turing machine accepting the language of words w representing 1 in $\mathcal{P}$ in non-deterministic time equivalent to the Dehn function of $\mathcal{P}$ and in non-deterministic space equivalent to the filling length function of $\mathcal{P}$.*

Sketch proof. The construction of a non-deterministic Turing machine we give here is based on [95], however we shirk the formal descriptions of machine states and transitions (which are set out carefully in [95]). Let $\langle X \mid R \rangle$ be a finite presentation of $\mathcal{P}$ where for all $r \in R$ we find R also contains all cyclic permutations of r and r^{-1}. Our machine has two tapes and two heads; its alphabet is $X^{\pm 1}$ and it will start with an input word w written on the first tape and with the second tape blank. At any time during the running of the machine there will be a word (possibly empty, but containing no blank spaces) on each tape and the remainder of the tapes will be blank; the head on each tape will always be located between two squares, that to its right being blank and that to its left containing the final letter of the word on the tape (assuming that word is not empty). We allow a head to see only the one square to its right and the $\rho := \max_{r \in R} \ell(r)$ squares to its left (a convenience to make the *apply-a-relator* transition reversible). The transitions are:

Shift. Erase the final letter x from the word on the end of one tape and insert x^{-1} on the end of the word on the other tape.

Apply-a-relator. Append/remove a word from R to/from the word on the first tape as a suffix.

Expand/Reduce. Append/remove a letter in $X^{\pm 1}$ to/from both words.

Accept. Accept w if the squares seen by the two heads are empty.

    Notice that acceptance occurs precisely when both tapes are empty. Consider the words one gets by concatenating the word on the first tape with the inverse of the word on the second at any given time — as the machine progresses this gives a null-sequence; indeed, (modulo the fact that it may take multiple transitions to realise each move) any null-sequence can be realised as a run of the machine. So, as the SPACE of a run of the machine is the maximum number of non-blank squares in the course of the run, the space complexity of this machine is equivalent to the filling length function of $\mathcal{P}$.

    Checking that the time complexity of our machine is equivalent to the Dehn function of $\mathcal{P}$ requires more care. Suppose w is a null-homotopic word. It is clear that any run of the machine on input w takes time at least $\text{Area}(w)$ because it must involve at least $\text{Area}(w)$ *apply-a-relator* transitions. For the reverse bound we wish to show that there is a run of the machine on input w in which not only the number of *apply-a-relator* transitions, but also the number of *shift* and *expand/reduce* transitions, can be bounded above by a constant times $\text{Area}(w) + \ell(w)$. Sapir [94, Section 3.2] suggests using [82, Lemma 1] to simplify the proof of [95]. We outline a similar approach based on the proof of Theorem 2.2.3.

    Let Δ be a minimal area van Kampen diagram for w and let T be a spanning tree in the 1-skeleton of Δ. Note that the total number of edges in T is at most

$\rho \operatorname{Area}(w) + \ell(w)$. A shelling of Δ is described in the proof of Theorem 2.2.3 in which one collapses 1-cells and 2-cells as one encounters them when following the boundary circuit of a small neighbourhood of T. Realise the null-sequence of this shelling (that is the sequence of boundary words of its diagrams), as the run of our Turing machine. The total number of moves is controlled as required because the shift moves correspond to the journey around T. □

The message of these results is that $\operatorname{Area} : \mathbb{N} \to \mathbb{N}$ and $\operatorname{FL} : \mathbb{N} \to \mathbb{N}$ are complexity measures of the word problem for $\mathcal{P}$ — they are the NON-DETERMINISTIC TIME (up to $\simeq$-equivalence) and the NON-DETERMINISTIC SPACE of the naïve approach to solving the word problem in the Dehn proof system by haphazardly applying relators, freely reducing and freely expanding in the hope of achieving an empty tape. This theme is taken up by Birget [11] who describes it as a "connection between *static* fillings (like length and area) and *dynamic* fillings (e.g., space and time complexity of calculations)." It opens up the tantalising possibility of, given a computational problem, translating it, suitably effectively, to the word problem in the Dehn proof system of some finite presentation. Then geometric considerations can be brought to bear on questions concerning algorithmic complexity.

Striking results in this direction have been obtained by Sapir and his collaborators using $\mathcal{S}$-*machines*. See [94] for a recent survey. Their results include the following.

Theorem 1.5.5 (Sapir–Birget–Rips [95]). *If $f(n)$ is the time function of a nondeterministic Turing machine such that $f^4(n)$ is super-additive (that is, $f^4(m + n) \geq f^4(m) + f^4(n)$ for all m, n), then there is a finite presentation with $\operatorname{Area}(n) \simeq f^4(n)$ and $\operatorname{IDiam}(n) \simeq f^3(n)$.*

Building on [10, 77, 95], a far-reaching *controlled* embedding result (Theorem 1.1) is proved in [12], of which the following theorem is a remarkable instance. Theorem 1.5.4 gives the *if* implication; the hard work is in the converse. [Recall that a problem is NP when it is decidable in non-deterministic polynomial time.]

Theorem 1.5.6 (Birget–Ol'shanskii–Rips–Sapir [12]). *The word problem of a finitely generated group G is an NP-problem if and only if $G \hookrightarrow \hat{G}$ for some finitely presentable group $\hat{G}$ such that the Dehn function of $\hat{G}$ is bounded above by a polynomial. Indeed, $G \hookrightarrow \hat{G}$ can be taken to be a quasi-isometric embedding.*

Birget conjectures the following analogue of Theorem 1.5.6 for filling length, relating it to NON-DETERMINISTIC SYMMETRIC SPACE complexity. The *if* part of the conjecture is covered by Theorem 1.5.4.

Conjecture 1.5.7 (Birget [11]). For a finitely generated group G and function $f : \mathbb{N} \to \mathbb{N}$, there is a non-deterministic symmetric Turing machine accepting the language of words w representing 1 in G within space $\preceq f(\ell(w))$ if and only if G embeds in a finitely presentable group with filling length function $\preceq f$.

Another approach is pursued by Carbone in [26, 27] where she relates van Kampen diagrams to *resolution proofs* in logic, in such a way that the Dehn function corresponds to proof length.

Our discussions so far in this section get us some way towards a proof of the following theorem. We encourage the reader to complete the proof as an exercise (or to refer to [47] or Part I of this volume).

Theorem 1.5.8 (Gersten [47]). *For a finite presentation the following are equivalent.*

- *The word problem for $\mathcal{P}$ is solvable.*

- Area $: \mathbb{N} \to \mathbb{N}$ *is bounded above by a recursive function.*

- Area $: \mathbb{N} \to \mathbb{N}$ *is a recursive function.*

So, strikingly, finite presentations of groups with unsolvable word problem (which do exist [13, 73]) have Dehn functions out-growing all recursive functions. Moreover, with the results we will explain in Chapter 2, it can be shown that for a finite presentation, *one* of the filling functions

$$\text{Area}, \text{FL}, \text{IDiam}, \text{GL}, \text{DGL} : \mathbb{N} \to \mathbb{N}$$

is a recursive function if and only if *all* are recursive functions.

However, in general, Dehn function is a poor measure of the time-complexity of the word problem for a group G. Cohen [30] and Madlener & Otto [68] showed that amongst all finite presentations of groups, there is no upper bound on the size of the gap between the Dehn function and the time-complexity of the word problem (in the sense of the Grzegorczyk hierarchy).

Indeed [94], the constructions of [12, 95] can be used to produce finite presentations with word problems solvable in quadratic time but arbitrarily large (recursive) Dehn functions. The idea is that an embedding of a finitely generated group G into another group $\hat{G}$ that has an efficient algorithm to solve its word problem leads to an efficient algorithm for the word problem in G. For example, $\langle a, b \mid a^b = a^2 \rangle$ has an exponential Dehn function (Proposition 2.1.6), but a polynomial time word problem:

Exercise 1.5.9. Find a deterministic polynomial time algorithm for $\langle a, b \mid a^b = a^2 \rangle$. Hint: $\langle a, b \mid a^b = a^2 \rangle$ is subgroup of $\text{GL}_2(\mathbb{Q})$ via $a = \begin{pmatrix} 1 & 1 \\ 0 & 1 \end{pmatrix}$ and $b = \begin{pmatrix} 1/2 & 0 \\ 0 & 1 \end{pmatrix}$.

In fact, this approach yields an $n(\log n)^2(\log \log n)$ time solution — see [12].

The group

$$\langle a, b \mid a^{a^b} a^{-2} \rangle,$$

introduced by Baumslag in [4], is an even more striking example. Its Dehn function was identified by Platonov [90] as

$$\text{Area}(n) \simeq \overbrace{f(f \ldots (f(1)) \ldots)}^{\log_2 n},$$

where $f(n) := 2^n$. Earlier Gersten [45, 47] had shown it's Dehn function to grow faster than every iterated exponential and Bernasconi [8] had found a weaker upper bound. However I. Kapovich & Schupp [96] and Miasnikov, Ushakov & Wong [69] claim its word problem is solvable in polynomial time. In the light of this and Magnus' result that every 1-relator group has solvable word problem [67], Schupp sets the challenge [96]:

Open Problem 1.5.10. Find a one-relator group for which there is, provably, no algorithm to solve the word problem within linear time.

In this context we mention Bernasconi's result [8] that all Dehn functions of 1-relator groups are at most (that is, $\preceq$) Ackermann's function (which is recursive but not primitive recursive), and a question of Gersten:

Open Problem 1.5.11. Is the Dehn function of every one-relator group bounded above by (that is, $\preceq$) the Dehn function of $\langle\, a, b \mid a^{a^b} a^{-2} \,\rangle$?

1.6   Filling functions for Riemannian manifolds

For a closed, connected Riemannian manifold M, define $\text{Area}_M : [0, \infty) \to [0, \infty)$ by

$$\text{Area}_M(l) := \sup_{c} \inf_{D} \{\, \text{Area}(D) \mid D : \mathbb{D}^2 \to \widetilde{M}, \, D|_{\partial \mathbb{D}^2} = c, \, \ell(c) \leq l \,\},$$

where $\text{Area}(D)$ is 2-dimensional Hausdorff measure and the infimum is over all Lipschitz maps D. Some remarks, following Bridson [18]: by Morrey's solution to Plateau's problem [71], for a fixed c, the infimum is realised; and, due to the regularity of the situation, other standard notions of area would be equivalent here.

One can construct a Γ-equivariant map Φ of $Cay^2(\mathcal{P})$ into the universal cover $\widetilde{M}$ of M by first mapping 1 to some base point v, then extending to Γ (which we identify with the 0-skeleton of $Cay^2(\mathcal{P})$) by mapping $\gamma \in \Gamma$ to the translate of v under the corresponding deck transformation, then extending to the 1-skeleton by joining the images of adjacent vertices by an equivariantly chosen geodesic for each edge, and then to the 2-skeleton by equivariantly chosen finite area fillings for each 2-cell. Loops in M can be approximated by *group-like paths* — the images under Φ of edge-paths in $Cay^2(\mathcal{P})$ — and maps of discs $D : \mathbb{D}^2 \to \widetilde{M}$ can be related to images under Φ of van Kampen diagrams into $\widetilde{M}$ (or, more strictly, compositions with Φ). This is the intuition for the following result, however a full proof requires considerable technical care.

Theorem 1.6.1 (The Filling Theorem [18, 25, 59]). *For a closed, connected Riemannian manifold M, the Dehn function of any finite presentation of $\Gamma := \pi_1 M$ is $\simeq$-equivalent to* $\mathrm{Area}_M : [0, \infty) \to [0, \infty)$.

Similar results hold for intrinsic diameter, extrinsic diameter and filling length. The *intrinsic diameter* of a continuous map $D : \mathbb{D}^2 \to \widetilde{M}$ is

$$\mathrm{IDiam}_M(D) := \sup \{ \rho(a, b) \mid a, b \in \mathbb{D}^2 \}$$

where ρ is the pull back of the Riemannian metric:

$$\rho(a, b) = \inf \{ \ell(D \circ p) \mid p \text{ a path in } \mathbb{D}^2 \text{ from } a \text{ to } b \}.$$

[Note that there may be distinct points $a, b \in \mathbb{D}^2$ for which $\rho(a, b) = 0$ — that is, ρ may be only a *pseudo*-metric $\mathbb{D}^2$.] The extrinsic diameter $\mathrm{EDiam}_M(D)$ is the diameter of $D(\mathbb{D}^2)$, as measured with the distance function on $\widetilde{M}$. These two notions of diameter give functionals on the space of rectifiable loops in $\widetilde{M}$: the *intrinsic* and *extrinsic* diameter functionals $\mathrm{IDiam}_M, \mathrm{EDiam}_M : [0, \infty) \to [0, \infty)$ — at l these functionals take the values

$$\sup_c \inf_{D \in \mathcal{D}} \{ \mathrm{IDiam}_M(D) \mid D : \mathbb{D}^2 \to \widetilde{M}, D \mid_{\partial \mathbb{D}^2} = c, \ell(c) \leq l \} \quad \text{and}$$

$$\sup_c \inf_{D \in \mathcal{D}} \{ \mathrm{EDiam}_M(D) \mid D : \mathbb{D}^2 \to \widetilde{M}, D \mid_{\partial \mathbb{D}^2} = c, \ell(c) \leq l \},$$

respectively.

The filling length $\mathrm{FL}(c)$ of a rectifiable loop $c : [0, 1] \to \widetilde{M}$ is defined in [61] to be the infimal length L such that there is a null-homotopy $H : [0, 1]^2 \to \widetilde{M}$ where $H(s, 0) = c(s)$ and $H(0, t) = H(1, t) = H(s, 1)$ for all $s, t \in [0, 1]$, such that for all $t \in [0, 1]$ we find $s \mapsto H(s, t)$ is a loop of length at most L. Then the *filling length function* FL_M is defined to be the supremum of $\mathrm{FL}(c)$ over all rectifiable loops $c : [0, 1] \to \widetilde{M}$ of length at most c.

The analogue of the Filling Theorem is:

Theorem 1.6.2. [22, 23] *If $\mathcal{P}$ is a finite presentation of the fundamental group Γ of a closed, connected Riemannian manifold M, then*

$$\mathrm{IDiam}_\mathcal{P} \simeq \mathrm{IDiam}_M, \quad \mathrm{EDiam}_\mathcal{P} \simeq \mathrm{EDiam}_M, \quad \text{and} \quad \mathrm{FL}_\mathcal{P} \simeq \mathrm{FL}_M.$$

1.7   Quasi-isometry invariance

The Dehn function is a *presentation invariant* and, more generally, a *quasi-isometry* invariant, in the sense of the following theorem. For a proof and background see [2] or [97, Theorem 4.7].

Theorem 1.7.1. *If G, H are groups with finite generating sets X, Y, respectively, if (G, d_X) and (H, d_Y) are quasi-isometric, and if G is finitely presentable, then H is finitely presentable. Moreover, if G and H have finite presentations $\mathcal{P}$ and $\mathcal{Q}$, respectively, then the associated Dehn functions $\mathrm{Area}_{\mathcal{P}}, \mathrm{Area}_{\mathcal{Q}} : \mathbb{N} \to \mathbb{N}$ satisfy $\mathrm{Area}_{\mathcal{P}} \simeq \mathrm{Area}_{\mathcal{Q}}$. In particular, the Dehn functions of two finite presentations of the same group are $\simeq$ equivalent.*

Thus it makes sense to say that a finitely presentable group G has a linear, quadratic, n^{α} (with $\alpha \geq 1$), exponential etc. Dehn function, meaning

$$\mathrm{Area}(n) \simeq n, n^2, n^{\alpha}, \exp(n), \ldots$$

with respect to some, and hence any, finite presentation. Note also, that for $\alpha, \beta \geq 1$ and $f, g : [0, \infty) \to [0, \infty)$ defined by $f(n) = n^{\alpha}$ and $g(n) = n^{\beta}$ we have $f \simeq g$ if and only if $\alpha = \beta$.

Results analogous to Theorem 1.7.1 also apply to EDiam, IDiam and FL — see [22, 23]. In each case the proof involves monitoring diagram measurements is the course of the standard proof that finite presentability is a quasi-isometry invariant — the first such quantitative version was [2]. Proofs of the independence, up to $\simeq$ equivalence, of IDiam : $\mathbb{N} \to \mathbb{N}$ and FL : $\mathbb{N} \to \mathbb{N}$ on the finite presentation predating [22, 23] use Tietze transformations and are in [55] and [52]. In [51] it is shown that the gallery length functions of two *fat* finite presentations of the same group are $\simeq$ equivalent. One expects that *fat* finite presentations of quasi-isometric groups would have $\simeq$ equivalent gallery length functions, and DGL functions, likewise.

Table 1.1: A summary of the multiple interpretations of filling functions

	van Kampen diagrams $\pi : \Delta \to Cay^2(\mathcal{P})$ for w	$P = \prod_{i=1}^{N} u_i^{-1} r_i^{\epsilon_i} u_i$ equalling w in $F(X)$	Dehn proof system	$G = \pi_1(M)$
Area(w)	$\min_{\Delta} \{\# \text{ of 2-cells in } \Delta\}$	$\min_{P} N$	NON-DETER-MINISTIC TIME	Riemannian Area
FL(w)	$\min\{ \text{FL}(\mathcal{S}) \mid \text{shellings } \mathcal{S} = (\Delta_i) \text{ of } \Delta \}$, where $\text{FL}(\mathcal{S}) = \max_i \ell(\partial \Delta_i)$	?	SPACE	Riemannian FL
IDiam(w)	$\min_{\Delta} \text{Diam } \Delta^{(1)}$	$\min_{P} \max_i \ell(u_i)$	?*	Riemannian IDiam
EDiam(w)	$\min_{\Delta} \text{Diam } \pi(\Delta^{(1)})$	$\min_{P} \max \{ d_X(\star, u) \mid u \text{ a prefix of } w\}$	?	Riemannian EDiam
GL(w)	$\min_{\Delta} \text{Diam } (\Delta^*)^{(1)}$	?	?	?
DGL(w)	$\min_{\Delta}\{\text{Diam}T + \text{Diam}T^* \mid T \text{ a spanning tree in } \Delta^{(1)}\}$	?	?**	?**

It remains a challenge to supply appropriate interpretations (if they exist) in place of each question mark (modulo * and **).

* Birget [11, Proposition 5.1] relates IDiam(w) to an exponential of the deterministic time of an approach to solving the word problem by constructing a non-deterministic finite automata based on part of the Cayley graph.

** By Theorem 2.2.6 we have DGL $\simeq$ FL for *fat* presentations.

Chapter 2

Relationships Between Filling Functions

This chapter concerns relationships between filling functions that apply irrespective of the group being presented. A more comprehensive account of known relationships is in [54], along with a description of how there are many additional coincidences between filling functions if one only uses van Kampen diagrams whose vertices have valence at most some constant.

We fix a finite presentation $\mathcal{P} = \langle X \mid R \rangle$ of a group G for the whole chapter. Some relationships are easy to find. Suppose $\pi : \Delta \to Cay^2(\mathcal{P})$ is a van Kampen diagram with base vertex $\star$. Then

$$\mathrm{IDiam}(\Delta) \ \leq \ \mathrm{FL}(\Delta)$$

because the images in Δ of the boundaries $\partial \Delta_i$ of the diagrams in a shelling form a family of contracting loops that at some stage pass through any given vertex v and so provide paths of length at most $\mathrm{FL}(\Delta)/2$ to $\star$. Defining $K := \max \{ \ell(r) \mid r \in R \}$, the total number of edges in Δ is at most $K\,\mathrm{Area}(\Delta) + \ell(\partial\Delta)$ and so

$$\mathrm{FL}(\Delta) \ \leq \ K\,\mathrm{Area}(\Delta) + \ell(\partial\Delta).$$

It follows that for all n,

$$\mathrm{IDiam}(n) \ \leq \ \mathrm{FL}(n) \ \leq \ K\,\mathrm{Area}(n) + n.$$

Similarly, one can show that

$$\mathrm{GL}(n) \ \leq \ \mathrm{DGL}(n) \ \leq \ K\,\mathrm{Area}(n) + n \quad \text{and} \quad \mathrm{IDiam}(n) \ \leq \ K\,\mathrm{GL}(n).$$

Next we give the *"space-time"* bound [52, Corollary 2], [61, 5.C] — an analogue of a result in complexity theory.

Proposition 2.0.2. *Define $K := 2\,|X| + 1$. Then for all n,*

$$\mathrm{Area}(n) \leq K^{\mathrm{FL}(n)}.$$

Proof. This result is most transparent in the Dehn proof system of Section 1.5. The number of words of length at most $FL(n)$, and hence the number of different words occurring in a null-sequence for a null-homotopic word w of length at most n, is at most $K^{FL(n)}$. Every such w has a null-sequence in which no word occurs twice and so there are at most $K^{FL(n)}$ application-of-a-relator moves in this null sequence. $\square$

Proposition 2.0.2 and Theorem 1.3.4 combine to give the following result which may be surprising in that it makes no reference to the length function of the combing.

Corollary 2.0.3. [48, 52] *The Dehn function of an (a)synchronously combable group grows at most exponentially fast.*

2.1   The Double Exponential Theorem

Gallery length can be used to prove a theorem of D.E.Cohen [29], known as *the Double Exponential Theorem.* Cohen's proof involves an analysis of the Nielsen reduction process. A proof, using *Stallings folds,* was given by Gersten [44] and was generalised by Papasoglu [88] to the more general setting of filling edge-loops in a simply connected complex of uniformly bounded local geometry. Birget found a proof based on context free languages [11]. The proof below is from [51]. It combines two propositions each of which establish (at most) exponential leaps, the first from IDiam up to GL, and the second from GL up to Area.

A key concept used is that of a *geodesic spanning tree* based at a vertex $\star$ in a graph Γ — that is, a spanning tree such that for every vertex v in Γ, the combinatorial distances from v to $\star$ in Γ and in T agree.

Exercise 2.1.1. Prove that for every finite connected graph Γ and every vertex $\star$ in Γ there is a geodesic spanning tree in Γ based at $\star$.

Theorem 2.1.2. *There exists $C > 0$, depending only on $\mathcal{P} = \langle X \mid R \rangle$, such that*

$$\text{Area}(n) \ \leq \ n\,C^{C^{\text{IDiam}(n)}}$$

for all $n \in \mathbb{N}$.

Define the *based* intrinsic diameter of a diagram Δ with base vertex $\star$ by

$$\text{IDiam}_\star(\Delta) \ := \ \max\Big\{ \ \rho(\star, v) \ \mid \ v \in \Delta^{(0)} \ \Big\},$$

where ρ denotes the combinatorial metric on $\Delta^{(1)}$. Note that, by definition, $\text{IDiam}_\star(\Delta) \leq \text{IDiam}(\Delta)$.

Proposition 2.1.3. *Suppose w is a null-homotopic word and Δ is a van Kampen diagram for w such that $\mathrm{IDiam}_\star(\Delta)$ is minimal. Moreover, assume Δ is of minimal area amongst all such diagrams. For $A := 2|X| + 1$,*

$$\mathrm{GL}(\Delta) \;\leq\; 2A^{1 + 2\mathrm{IDiam}(\Delta)}.$$

So $\mathrm{GL}(n) \leq 2A^{1 + 2\mathrm{IDiam}(n)}$ for all n.

Proof. Let T be a geodesic spanning tree in $\Delta^{(1)}$ based at $\star$. (See Figure 2.1.) To every edge e in $\Delta^{(1)} \smallsetminus T$ associate an anticlockwise edge-circuit γ_e in $\Delta^{(1)}$, based at $\star$, by connecting the vertices of e to $\Delta^{(1)}$ by geodesic paths in T. Let w_e be the word one reads along γ_e. Then $\ell(w_e) \leq 1 + 2\mathrm{IDiam}(\Delta^{(1)})$.

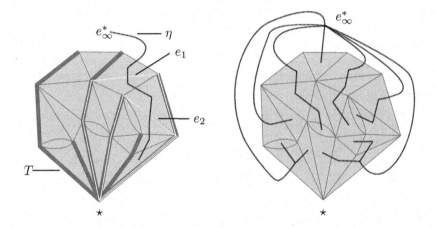

Figure 2.1: Illustrations of the proofs of Propositions 2.1.3 and 2.1.4.

Let T^* be the dual tree to T (as defined in Section 1.2) and e_∞^* be the vertex of T^* dual to the face e_∞. Suppose that e_1 and e_2 are two distinct edges of $\Delta^{(1)} \smallsetminus T$ dual to edges e_1^* and e_2^* of T^* on some geodesic η in T^* from e_∞^* to a leaf and that e_2^* is further from e_∞^* than e_1^* along η. Then γ_{e_1} and γ_{e_2} bound subcomplexes $C(\gamma_{e_1})$ and $C(\gamma_{e_2})$ of Δ in such a way that $C(\gamma_{e_2})$ is a subcomplex of $C(\gamma_{e_1})$. Suppose that w_{e_1} and w_{e_2} are the same words. Then we could cut $C(\gamma_{e_1})$ out of Δ and glue $C(\gamma_{e_2})$ in its place, producing a new van Kampen diagram for w with strictly smaller area and no increase in $\mathrm{IDiam}_\star$ — a contradiction.

Therefore $\ell(\eta) \leq A^{1 + 2\mathrm{IDiam}(\Delta)}$, as the right-hand side is an upper bound on the number of distinct words of length at most $1 + 2\mathrm{IDiam}(\Delta)$, and the result follows. $\qquad\square$

Proposition 2.1.4. *If Δ is a van Kampen diagram for a null-homotopic word w, then*

$$\text{Area}(\Delta) \leq \ell(w)(B+1)^{\text{GL}(\Delta)},$$

where $B := \max\{\ell(r) \mid r \in R\}$. It follows that $\text{Area}(n) \leq n(B+1)^{\text{GL}(n)}$ for all n.

Proof. Sum over $1 \leq k \leq \text{GL}(\Delta)$, the geometric series whose k-th term $n(B+1)^{k-1}$ dominates the number of vertices at distance k from e_∞^* in a geodesic spanning tree in the 1-skeleton of Δ^*, based at e_∞^*, as illustrated in Figure 2.1. $\qquad \square$

Remark 2.1.5. In fact, the two propositions above establish more than claimed in Theorem 2.1.2. They show that there exists $C > 0$, depending only on $\mathcal{P}$, such the bounds

$$\text{Area}(n) \; \leq \; n\,C^{C^{\text{Diam}(n)}} \quad \text{and} \quad \text{GL}(n) \; \leq \; C^{\text{Diam}(n)}$$

are simultaneously realisable on van Kampen diagrams that are of minimal IDiam amongst all diagrams for their boundary words. This together with an inequality of [52] show (as observed by Birget [11]) that $\text{IDiam}(n)$ is always no more than a single exponential of $\text{FL}(n)$.

Proposition 2.1.6. *The presentation $\langle\, a, b \mid a^b = a^2 \,\rangle$ has filling length and Dehn functions satisfying*

$$\text{FL}(n) \simeq n \quad \text{and} \quad \text{Area}(n) \simeq \exp(n).$$

Proof. That $\text{FL}(n) \preceq n$ follows from asynchronous combability and Theorem 1.3.4, and so $\text{Area}(n) \preceq \exp(n)$ by Proposition 2.0.2; $\text{FL}(n) \succeq n$ by definition of $\succeq$, and $\text{Area}(n) \succeq \exp(n)$ by the following lemma (see e.g. [53, Section 2.3], [61, Section 4C$_2$]) applied to fillings for words $[a, a^{b^n}]$ as illustrated in Figure 1.7 in the case $n = 6$. $\qquad \square$

A presentation $\mathcal{P}$ is *aspherical* when $Cay^2(\mathcal{P})$ is contractible.

Lemma 2.1.7 (Gersten's Asphericity Lemma [18, 46]). *If $\mathcal{P}$ is a finite aspherical presentation and $\pi : \Delta \to Cay^2(\mathcal{P})$ is a van Kampen diagram such that π is injective on the complement of $\Delta^{(1)}$, then Δ is of minimal area amongst all van Kampen diagrams with the same boundary circuit.*

Exercise 2.1.8. Give another proof that $\text{Area}(n) \succeq \exp(n)$ for $\langle\, a, b \mid a^b = a^2 \,\rangle$ by studying the geometry of b-*corridors* and the words in $\{a^{\pm 1}\}^*$ along their sides in van Kampen diagrams for $[a, a^{b^n}]$. (A b-*corridor* is a concatenation of 2-cells, each joined to the next along an edge labelled by b, and running between two oppositely oriented edges on $\partial\Delta$ both labelled b. For example, every horizontal strip of 2-cells in the van Kampen diagram in Figure 1.7 is a b-corridor.)

Remarkably, no presentation is known for which the leap from IDiam(n) to Area(n) is more than the single exponential of $\langle a, b \mid a^b = a^2 \rangle$. So many people (e.g. in print [29, 47, 61], but it is often attributed to Stallings and indeed said that he conjectured a positive answer) have raised the natural question:

Open Question 2.1.9. Given a finite presentation, does there always exist $C > 0$ such that Area(n) $\leq C^{\text{IDiam}(n)} + n$ for all n?

Indeed, one can ask more [51]: is a bound Area(n) $\leq C^{\text{IDiam}(n)} + n$ always realisable on van Kampen diagrams of minimal IDiam for their boundary words.

Gromov [61, page 100] asks:

Open Question 2.1.10. Given a finite presentation, does there always exist $C > 0$ such that FL(n) $\leq C(\text{IDiam}(n) + n)$ for all n?

This question has also been attributed to Casson [61, page 101]. Gromov notes that an affirmative answer to 2.1.10 would imply an affirmative answer to 2.1.9 by Proposition 2.0.2. At the level of combinatorial or metric discs the analogue of 2.1.10 is false — Frankel & Katz [43] produced a family of metric discs D_n with diameter and perimeter 1 but such that however one null-homotopes ∂D_n across D_n, one will (at some time) encounter a loop of length at least n; we will describe combinatorial analogues of these discs in Section 2.2.

In [51] Gersten and the author ask 2.1.10 with FL is replaced by GL:

Open Question 2.1.11. Given a finite presentation, does there always exist $C > 0$ such that GL(n) $\leq C(\text{IDiam}(n) + n)$ for all n?

As mentioned in Remark 2.1.5, like filling length, gallery length sits at most an exponential below the Dehn function and at most an exponential above the intrinsic diameter function, in general, and a positive answer to 2.1.11 would resolve 2.1.9 affirmatively. However, 2.1.11 may represent a different challenge to 2.1.10 because separating the GL and IDiam functions has to involve a family of van Kampen diagrams for which there is no uniform upper bound on the valence of vertices. The features of combinatorial discs Δ_n discussed in Section 2.2 may be reproducible in van Kampen diagrams in some group in such a way as to answer 2.1.10 negatively, but those discs have uniformly bounded vertex valences and so will not be of use for 2.1.11.

We now turn to what can be said in the direction of an affirmative answer to 2.1.9. For a null-homotopic word w, define

$$\overline{\text{IDiam}}(w) := \max\{ \text{IDiam}(\Delta) \mid \text{minimal area van Kampen diagrams } \Delta \text{ for } w \}.$$

Then, as usual, define a function $\mathbb{N} \to \mathbb{N}$, the *upper intrinsic diameter function*, by

$$\overline{\text{IDiam}}(n) := \max\{ \overline{\text{IDiam}}(w) \mid \text{words } w \text{ with } \ell(w) \leq n \text{ and } w = 1 \text{ in } G \}.$$

A reason why 2.1.9 is a hard problem is that, in general, minimal area and minimal diameter fillings may fail to be realisable simultaneously as illustrated by the example in the following exercise.

Exercise 2.1.12. By finding embedded van Kampen diagrams Δ_n for the words $w_n := \left[x^{t^n}, y^{s^n}\right]$ and applying Lemma 2.1.7, show that the Dehn and $\overline{\text{IDiam}}$ functions of the aspherical presentation

$$\langle\ x, y, s, t\ |\ [x, y],\ x^t x^{-2},\ y^s y^{-2}\ \rangle$$

of Bridson [16, 53] satisfy

$$\text{Area}(n)\ \succeq\ 2^n \quad\text{and}\quad \overline{\text{IDiam}}(n)\ \succeq\ 2^n.$$

But show that $\text{IDiam}(w_n) \preceq n$ for all n. (*Hint.* Insert many copies of suitable van Kampen diagrams for $x^{t^n} x^{-2^n}$ into Δ_n. In fact, it is proved in [53] that $\text{IDiam}(n) \preceq n$.)

Using $\overline{\text{IDiam}}$ side-steps the problem of minimal area and diameter bounds not being realised on the same van Kampen diagram and the following single exponential bound on Area can be obtained.

Theorem 2.1.13. [53] *Given a finite presentation* $\mathcal{P}$, *there exists* $K > 0$ *such that*

$$\text{Area}(n)\ \leq\ nK^{\overline{\text{IDiam}}(n)}.$$

Sketch proof. Suppose Δ is a minimal area van Kampen diagram for w. Recall that topologically Δ is a singular 2-disc and so is a tree-like arrangement of arcs and topological 2-discs. Consider successive *star-neighbourhoods* $\text{Star}^i(\partial\Delta)$ of $\partial\Delta$ defined by setting $\text{Star}^0(\partial\Delta)$ to be $\partial\Delta$ and $\text{Star}^{i+1}(\partial\Delta)$ to be the union closed cells that share at least a vertex with $\text{Star}^i(\partial\Delta)$. Define $A_i := \text{Star}^{i+1}(\partial\Delta)\smallsetminus\text{Star}^i(\partial\Delta)$, a sequence of *annuli* that partition Δ into $(A_i)_{i=0}^m$ where $m \leq K_0\text{IDiam}(\Delta)$ for some constant K_0 depending only on $\mathcal{P}$. This is illustrated in Figure 2.2. (The A_i will not be topological annuli, in general.)

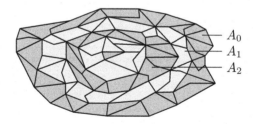

Figure 2.2: Successive star-neighbourhoods of the boundary of a diagram.

Suppose that for every vertex v on $\partial\Delta$ and every topological 2-disc component $\mathcal{C}$ of Δ, there are at most three edges in the interior of $\mathcal{C}$ that are all incident with v, all oriented away from or all towards v, and all have the same label. Then by considering 2-cells in $\mathcal{C}$ that are incident with v, we find there is constant K_1, depending only on $\mathcal{P}$, such that $\text{Area}(A_0)$ and the total length of the inner boundary circuit of A_0 are both at most $K_1\ell(\partial\Delta)$.

We then seek to argue similarly for $\Delta \smallsetminus A_0$, to get an upper bound on the length of the inner boundary and the area of A_1. And then continuing inductively at most $K_0 \mathrm{IDiam}(\Delta)$ times and summing an appropriate geometric series we will have our result.

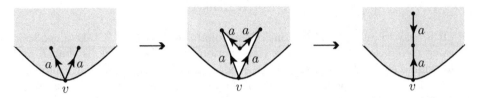

Figure 2.3: A diamond move.

The way we get this control on edges incident with vertices is to use *diamond moves*. These are performed on pairs of edges that are in the same 2-disc component of a van Kampen diagram, that are oriented towards or away from a common end-vertex $v \in \partial\Delta$, that have the same label, and whose other vertices, v_1 and v_2, are either both in the interior of Δ or both on $\partial\Delta$. The case where v_1 and v_2 are in the interior is illustrated in Figure 2.3. Performing a diamond moves when $v_1, v_2 \in \partial\Delta$ creates a new 2-disc component.

Do such diamond moves until none remain available — the process does terminate because diamond moves either increase the number of 2-disc components (which can only happen finitely many times) or decreases the valence of some boundary vertex without altering the number of 2-disc components. Then peal off A_0 and repeat the whole process on the boundary vertices of $\Delta \smallsetminus A_0$.

Continue likewise until the diagram has been exhausted. This takes at most $\sim \overline{\mathrm{IDiam}}(w)$ iterations because, whilst the diamond moves may change IDiam, they preserve Area. $\qquad\square$

Exercise 2.1.14. Use $\langle a, b \mid a^b = a^2 \rangle$ to show that the result of 2.1.13 is sharp, in general.

Open Problem 2.1.15. [53] Give a general upper bound for $\overline{\mathrm{IDiam}} : \mathbb{N} \to \mathbb{N}$ in terms of $\mathrm{IDiam} : \mathbb{N} \to \mathbb{N}$. Theorem 2.1 implies a double exponential bound holds; one might hope for a single exponential.

2.2 Filling length and duality of spanning trees in planar graphs

For a while, the following question and variants set out in [54], concerning planar graphs and dual pairs of trees in the sense of Section 1.2, constituted an impasse in the study of relationships between filling functions. [A *multigraph* is a graph in which edges can form loops and pairs of vertices can be joined by the multiple edges.]

Question 2.2.1. Does there exist $K > 0$ such that if Γ is a finite connected planar graph (or multigraph) then there is a spanning tree T in Γ such that

$$\mathrm{Diam}(T) \;\leq\; K\,\mathrm{Diam}(\Gamma) \quad \text{and}$$
$$\mathrm{Diam}(T^*) \;\leq\; K\,\mathrm{Diam}(\Gamma^*)\,?$$

It is easy to establish either one of the equalities with $K = 2$: take a geodesic spanning tree based at some vertex (see Section 2.1). The trouble is that T and T^* determine each other and they fight — altering one tree to reduce its diameter might increase the diameter of the other. The following exercise sets out a family of examples in which taking one of T and T^* to be a geodesic tree based at some vertex, does not lead to T and T^* having the properties of Question 2.2.1.

Exercise 2.2.2. Figure 2.4 shows the first four of a family of connected (multi-) graphs Γ_n. The horizontal path through Γ_n of length 2^n is a spanning tree T_n in Γ_n and T_n^* is a geodesic spanning tree in Γ_n^* based at the vertex dual to the face *at infinity*. Calculate $\mathrm{Diam}(T_n^*)$, $\mathrm{Diam}(\Gamma_n)$ and $\mathrm{Diam}(\Gamma_n^*)$. Find a spanning tree S_n in Γ_n such that $\mathrm{Diam}(S_n)$ and $\mathrm{Diam}(S_n^*)$ are both at most a constant times n.

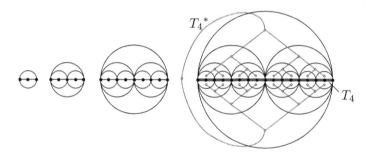

Figure 2.4: The graphs $\Gamma_1, \ldots, \Gamma_4$ of Exercise 2.2.2.

A *diagram* is a finite planar contractible 2-complex — in effect, a van Kampen diagram bereft of its group theoretic content. Question 2.2.1 can be regarded as a question about diagram measurements because a multigraph is finite, planar and connected if and only it is the 1-skeleton of a diagram. Question 2.2.1 was recently resolved negatively in [93] as we will explain. Filling length plays a key role via the following result (compare [61, 5.C]).

Theorem 2.2.3. *For all diagrams* Δ

$$\mathrm{FL}(\Delta) \;\leq\; C(\mathrm{DGL}(\Delta) + \ell(\partial\Delta)),$$

where C is a constant depending only on the maximum length of the boundary cycles of the 2-cells in Δ.

Proof (adapted from [51]). It suffices to show that given any spanning tree T in $\Delta^{(1)}$,

$$\mathrm{FL}(\Delta) \leq \mathrm{Diam}(T) + 2\lambda\,\mathrm{Diam}(T^*) + \ell(\partial\Delta), \tag{2.1}$$

where λ is the maximum length of the boundary circuits of 2-cells in Δ.

Let the base vertex $\star$ of Δ be the root of T and the vertex $\star_\infty$ dual to the *2-cell at infinity* be the root of T^*. Let m be the number of edges in T. Let $\gamma : [0, 2m] \to T$ be the edge-circuit in T that starts from $\star$ and traverses every edge of T twice, once in each direction, running close to the anticlockwise boundary loop of a small neighbourhood of T. For $i = 1, 2, \ldots, 2m$ let γ_i be the edge traversed by $\gamma|_{[i-1,i]}$ and consider it directed from $\gamma(i-1)$ to $\gamma(i)$. Let τ_i be the geodesic in T^* from $\star_\infty$ to the vertex v_i^* dual to the 2-cell to the right of γ_i (possibly $v_i^* = \star_\infty$). Let $\overline{\tau}_i$ be the union of all the 2-cells dual to vertices on τ_i.

There may be some 2-cells C in $\Delta \smallsetminus \bigcup_i \overline{\tau}_i$. For such a C, let e be the edge of $(\partial C) \smallsetminus T$ whose dual edge is closest in T^* to $\star_\infty$. Call such an edge e *stray*. Then e must be an edge-loop because otherwise the subdiagram with boundary circuit made up of e and the geodesics in T from the end vertices of e to $\star$ would contain some v_i^*. It follows that e is the boundary of a subdiagram that is of the form of Figure 2.5 save that shelling away a 2-cell reveals at most $(\lambda - 1)$ more 2-cells instead of two. We leave it to the reader to check that such a diagram has filling length at most $(\lambda - 1)\mathrm{Diam}(T^*)$.

Figure 2.5: Subdiagrams arising in the proof of Theorem 2.2.3.

To realise (2.1), shell the 2-cells of Δ in the following order: for $i = 1$, then $i = 2$, and so on, shell all the (remaining) 2-cells along $\overline{\tau}_i$, working from $\star_\infty$ to v_i^*. Except, whenever a *stray* edge e appears in the boundary circuit, pause and entirely shell away the diagram it contains. In the course of all the 2-cell collapses, do a *1-cell-collapse* whenever one is available.

One checks that in the course of this shelling, aside from detours into subdiagrams enclosed by stray edges, the anticlockwise boundary-circuit starting from $\star$ follows a geodesic path in T, then a path embedded in the 1-skeleton of some $\overline{\tau}_i$, before returning to $\star$ along $\partial\Delta$. These three portions of the circuit have lengths at most $\mathrm{Diam}(T)$, $\lambda\,\mathrm{Diam}(T^*)$ and $\ell(\partial\Delta)$, respectively. Adding $(\lambda - 1)\mathrm{Diam}(T^*)$ for the detour gives an estimate within the asserted bound. $\qquad\square$

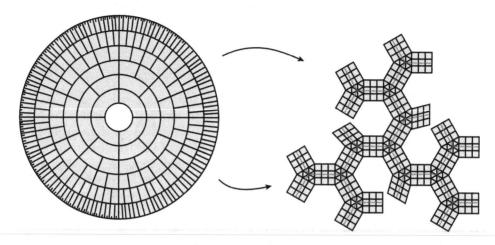

Figure 2.6: Attaching the outer boundary of the annulus to the boundary of the fattened tree gives the third of a family (Δ_n) of diagrams with filling length outgrowing diameter and dual diameter.

A family (Δ_n) of diagrams, essentially combinatorial analogues of metric 2-discs from [43], is constructed as follows to resolve Question 2.2.1. One inductively defines trees $\mathcal{T}_n$ by taking $\mathcal{T}_1$ to be a single edge and obtaining $\mathcal{T}_n$ from three copies of $\mathcal{T}_{n-1}$ by identifying a leaf vertex of each. (It is not important that this does not define $\mathcal{T}_n$ uniquely.) Fatten $\mathcal{T}_n$ to an n-thick diagram as shown in Figure 2.6 in the case $n = 3$. Then attach a combinatorial annulus around the fattened tree by identifying the outer boundary of the annulus (in the sense illustrated) with the boundary of the fattened tree. The annulus is made up of concentric rings, each with twice as many faces as the next, and is constructed in such a way that the length of its outer boundary circuit is the same as the length of the boundary circuit of the fattened tree; it serves as *hyperbolic skirt* bringing the diameters of $\Delta_n^{(1)}$ and its dual down to $\preceq n$.

The filling length of Δ_n grows $\preceq n^2$ — the reason being that a curve sweeping over $\mathcal{T}_n$ will, at some time, meet $n + 1$ different edges of $\mathcal{T}_n$; more precisely:

Lemma 2.2.4. *Suppose $\mathcal{T}_n$ is embedded in the unit disc $\mathbb{D}^2$. Let $\star$ be a basepoint on $\partial\mathbb{D}^2$. Suppose $H : [0,1]^2 \to \mathbb{D}^2$ is a homotopy satisfying $H(0,t) = H(1,t) = \star$ for all t, and $H_0(s) = e^{2\pi i s}$ and $H_1(s) = \star$ for all s, where H_t denotes the restriction of H to $[0,1] \times \{t\}$. Further, assume $H([0,1] \times [0,t]) \cap H([0,1] \times [t,1]) = H([0,1] \times \{t\})$ for all t. Then H_t meets at least $n + 1$ edges in $\mathcal{T}_n$ for some $t \in [0,1]$.*

The final details of the proof that (Δ_n) resolves Question 2.2.1 negatively are left to the following exercise.

Exercise 2.2.5.

1. Show that $\mathrm{Diam}(\Delta_n), \mathrm{GL}(\Delta_n) \preceq n$.

2. Prove Lemma 2.2.4 (*hint*: induct on n) and deduce that $\mathrm{FL}(n) \succeq n^2$.

3. Using Theorem 2.2.3, deduce a negative answer to Question 2.2.1.

The lengths of the boundary circuits in a $\mathcal{P}$-van Kampen diagram are at most the length of the longest defining relator in $\mathcal{P}$. So Theorem 2.2.3 shows that $\mathrm{FL} : \mathbb{N} \to \mathbb{N}$ and $\mathrm{DGL} : \mathbb{N} \to \mathbb{N}$ satisfy $\mathrm{FL} \preceq \mathrm{DGL}$. The reverse, $\mathrm{DGL} \preceq \mathrm{FL}$, is also true for *fat* presentations (defined in Section 1.3). The proof in [51] is technical but, roughly speaking, the idea is that a shelling of Δ gives a family of concentric edge-circuits in Δ contracting down to $\star$ which we can *fatten* and then have the arcs of both T and T^* follow these circuits. Thus we get

Theorem 2.2.6 ([51]). *The filling functions* GL, DGL *and* FL *for any finite fat presentation satisfy* $\mathrm{GL} \leq \mathrm{DGL} \simeq \mathrm{FL}$.

Despite the negative answer to Question 2.2.1, whether the inequality in this theorem can be replaced by a $\simeq$, rendering all three functions equivalent, remains an open question.

2.3 Extrinsic diameter versus intrinsic diameter

It is clear that for a van Kampen diagram $\pi : \Delta \to Cay^2(\mathcal{P})$ we have $\mathrm{EDiam}(\Delta) \leq \mathrm{IDiam}(\Delta)$ because paths in $\Delta^{(1)}$ are sent by π to paths in $Cay^2(\mathcal{P})$. But, in general, one expects a shortest path between $\pi(a)$ and $\pi(b)$ in the 1-skeleton of $Cay^2(\mathcal{P})$ to take a shortcut through the space and so not lift to a path in $\Delta^{(1)}$, so $\mathrm{EDiam}(\Delta)$ could be strictly less than $\mathrm{IDiam}(\Delta)$. In [22] finite presentations are constructed that confirm this intuition; indeed, the examples show that (multiplicative) polynomial gaps of arbitrarily large degree occur:

Theorem 2.3.1. [22] *Given* $\alpha > 0$, *there exists a finite presentation with*

$$n^\alpha \, \mathrm{EDiam}(n) \preceq \mathrm{IDiam}(n).$$

This result prompts the question of how far the gap can be extended:

Open Question 2.3.2. [22] What is the optimal upper bound for $\mathrm{IDiam} : \mathbb{N} \to \mathbb{N}$ in terms of $\mathrm{EDiam} : \mathbb{N} \to \mathbb{N}$, in general?

2.4 Free filling length

Recall from Section 1.2 that $\mathrm{FL}(\Delta)$ is defined with reference to a base vertex $\star$ on $\partial\Delta$. If we remove the requirement that $\star$ be preserved throughout a shelling then we get $\mathrm{FFL}(\Delta)$: the minimal L such that there is a *free* combinatorial null-homotopy of $\partial\Delta$ across Δ — see Figure 2.7. We then define $\mathrm{FFL}(w)$ for null-homotopic w, and the *free filling length function* $\mathrm{FFL} : \mathbb{N} \to \mathbb{N}$ in the usual way.

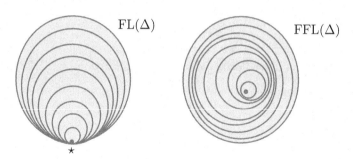

Figure 2.7: Base-point-fixed vs. free filling length.

Exercise 2.4.1. Re-interpret FFL(w) in terms of the Dehn Proof System. (*Hint.* Allow one extra type of move to occur in null-sequences.)

It is questionable whether the base point plays a significant role in the definition of FL:

Open Problem 2.4.2. [23] Does there exist a finite presentation for which FL(n) $\not\simeq$ FFL(n)?

However, the following example gives null-homotopic words w_n for which FL(w_n) and FFL(w_n) exhibit dramatically different qualitative behaviour. Nonethe-less the filling functions agree: FL(n) $\simeq$ FFL(n) $\simeq 2^n$. (Exercise 2.4.4 provides some hints towards proof of these remarks; details are in [23].)

Example 2.4.3. [23] Let G be the group with presentation

$$\langle\, a, b, t, T, \tau \mid b^{-1}aba^{-2}, [t, a], [\tau, at], [T, t], [\tau, T]\,\rangle$$

and define $w_n := \left[T, a^{-b^n}\tau a^{b^n}\right]$. Then w_n is null-homotopic, has length $8n + 8$, and FFL(w_n) $\simeq n$, but FL(w_n) $\simeq 2^n$.

Exercise 2.4.4.

1. Show that FL(w_n) $\simeq 2^n$. (*Hint.* The van Kampen diagram Δ_n for w_n shown in Figure 2.8 in the case $n = 3$, has 2^n concentric t-*rings* — concatenations of 2-cells forming an annular subdiagram whose internal edges are all labelled by t. Show that every van Kampen diagram for w_n has t-rings nested 2^n deep.)

2. Show that FFL(w_n) $\simeq n$. (*Hint.* Produce a diagram Δ'_n from Δ_n such that FFL(Δ'_n) $\preceq n$ by making cuts along paths labelled by powers of a and gluing in suitable diagrams over the subpresentation $\langle a, b \mid b^{-1}aba^{-2}\rangle$.)

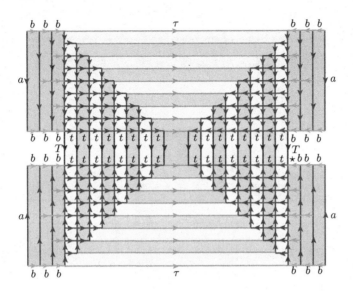

Figure 2.8: A van Kampen diagram Δ_3 for w_3 with concentric t-rings.

3. Find a family of words w'_n with $\mathrm{FFL}(w'_n) \simeq 2^n$ and $\ell(w'_n) \simeq n$.

Chapter 3

Example: Nilpotent Groups

Recall that a group G is nilpotent of class c if it has lower central series

$$G = \Gamma_1 \gtrapprox \Gamma_2 \gtrapprox \cdots \gtrapprox \Gamma_{c+1} = \{1\},$$

defined inductively by $\Gamma_1 := G$ and $\Gamma_{i+1} := [G, \Gamma_i]$.

3.1 The Dehn and filling length functions

The following theorem and its proof illustrate the value of two key ideas: studying not only Area but how it interacts with other filling functions, and using the *Dehn proof system* (see Section 1.5) to manipulate words and monitor the Dehn and filling length function.

Theorem 3.1.1. *If G is a finitely generated nilpotent group of class c, then the Dehn and filling length functions of G satisfy*

$$\mathrm{Area}(n) \preceq n^{c+1} \quad and \quad \mathrm{FL}(n) \preceq n.$$

Moreover, these bounds are simultaneously realisable.

The combinatorial proof in [50] is the culmination of a number of results: [31, 45, 64]. A more general result was proved by Gromov [61, 5.A_5'], [62] using masterful but formidable analytical techniques. The linear upper bound on filling length was proved prior to Theorem 3.1.1 by the author in [91] via asymptotic cones: finitely generated nilpotent groups have simply connected (indeed contractible) asymptotic cones by work of Pansu [83], and so the result follows as we will see in Theorem 4.3.3. Pittet [89], following Gromov, proved that a lattice in a simply connected homogeneous nilpotent Lie group of class c admits a polynomial isoperimetric function of degree $c+1$. (A nilpotent Lie group is called *homogeneous* if its Lie algebra is *graded*.)

No better general upper bound in terms of class is possible because, for example, the Dehn function of a free nilpotent group of class c is $\simeq n^{c+1}$ — see

Part I of this volume or [5, 47]. However, it is not best possible for individual nilpotent groups: the $2k + 1$ dimensional integral Heisenberg groups are of class 2 but have Dehn functions $\simeq n^2$ when $k > 1$ [1], [61, 5.A$_4'$], [81]. By way of contrast, the 3-dimensional integral Heisenberg group $\mathcal{H}_3$ is free nilpotent of class 2 and so has a cubic Dehn function.

A full proof of Theorem 3.1.1 is beyond the scope of this text. We will illustrate the ideas in the proof in [50] in the case of $\mathcal{H}_3$, presented by

$$\mathcal{P} := \langle\, x, y, z \mid [x, y]z^{-1}, [x, z], [y, z] \,\rangle.$$

The proof is via the Dehn proof system. We will show that there is a sequence (w_i) that uses $\preceq n^{c+1}$ application-of-a-relator moves and has $\max_i \ell(w_i) \preceq n$. This suffices — see Section 1.5.

The following lemma gives a means of *compressing* powers of z. It is a special case of [50, Corollary 3.2], which concerns non-identity elements in the final nontrivial group Γ_c in the lower central series. We use the terminology of Section 1.5. Fix $n \in \mathbb{N}$ (which in the proof will be the length of the word we seek to fill). For integers $s \geq 0$, define *compression words* $u(s)$ representing z^s in $\mathcal{H}_3$: if $0 \leq s \leq n^2 - 1$, then

$$u(s) \quad := \quad z^{s_0}[x^n, y^{s_1}],$$

where $s = s_0 + s_1 n$ for some $s_0, s_1 \in \{0, 1, \dots, n - 1\}$; define

$$u\left(n^2\right) \quad := \quad [x^n, y^n]$$

and

$$u\left(A + Bn^2\right) \quad := \quad u(A)\, u\left(n^2\right)^B$$

for all integers A, B with $0 \leq A \leq n^2 - 1$ and $B > 0$. Note that $\ell(u(A + Bn^2)) \leq K_0 n$ where K_0 depends only on B.

Lemma 3.1.2. *Fix $K_1 > 0$. There exists $K_2 > 0$ such that for all integers $K_1 n^2 \geq s \geq 0$, there is a concatenation of $\mathcal{P}$-sequences*

$$z^s \;\rightarrow\; z^s\, u(0) \;\rightarrow\; z^{s-1}\, u(1) \;\rightarrow\; z^{s-2}\, u(2) \;\rightarrow\; \cdots \;\rightarrow\; z\, u(s-1) \;\rightarrow\; u(s),$$

that converts z^s to $u(s)$, and uses at most $K_2 n^3$ application-of-a-relator moves.

We leave the proof of the lemma as an exercise except to say that the key is absorbing z^n subwords into commutators $[x^n, y^k]$ in which $k < n$, via:

$$z^n\left[x^n, y^k\right] \;=\; z^n x^{-n} y^{-k} x^n y^k \;=\; z^n x^{-n} y^{-k} y^{-1} y x^n y^k$$
$$=\; z^n x^{-n} y^{-(k+1)} (xz^{-1})^n y^{k+1} \;=\; \left[x^n, y^{k+1}\right].$$

A diagrammatic understanding of this calculation can be extracted from Figure 3.1 — the diagonal line running across the diagram from $\star$ is labelled by z^{16} and the half of the diagram below this diagonal demonstrates the equality $z^{16} = [x^4, y^4]$.

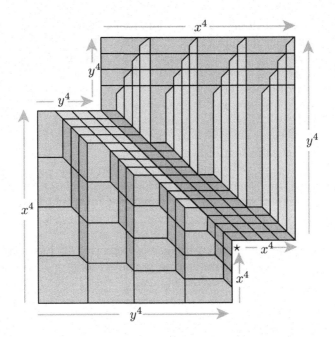

Figure 3.1: A van Kampen diagram for $[x^{-4}, y^{-4}][y^4, x^4]$ in the presentation $\langle x, y, z \mid [x, y]z^{-1}, [x, z], [y, z] \rangle$ for the 3-dimensional integral Heisenberg group $\mathcal{H}_3$. The diagonal line beginning at $\star$ is labelled by z^{16}.

The proof of Theorem 3.1.1 is an induction on the class. The base case of class 1 concerns finitely generated abelian groups G and is straightforward — see Example 1.5.3 (2). For the induction step one considers G/Γ_c, which is a finitely generated nilpotent group of class $c-1$. Let $\overline{\mathcal{P}} = \langle \overline{X} \mid \overline{R} \rangle$ be a finite presentation for G/Γ_c. Let $X = \overline{X} \cup X'$ where X' is a finite set of length c commutators that span Γ_c. The relators $\overline{R}$ represent elements of Γ_c in G, so there is a finite presentation for $\langle X \mid R \rangle$ in which the elements of R express the equality of elements of $\overline{R}$ with words on X'.

Given a length n null-homotopic word w in $\mathcal{P}$, the word $\overline{w}$ obtained from w by deleting all letters in X' is null-homotopic in $\overline{G}$ and so admits a null-$\overline{\mathcal{P}}$-sequence $(\overline{w}_i)_{i=0}^{\overline{m}}$ which, by induction, involves words of length $\preceq n$ and uses $\preceq n^c$ application-of-a-relator moves.

We seek to lift $(\overline{w}_i)_{i=0}^{\overline{m}}$ to a null-$\mathcal{P}$-sequence for w in which the words have length $\preceq n$ and $\preceq n^{c+1}$ application-of-a-relator moves are used. We illustrate how this works in the case of $G = \mathcal{H}_3$. We have

$$\overline{G} = \Gamma_1/\Gamma_2 = \mathcal{H}_3/\langle z \rangle,$$

which is presented by $\overline{\mathcal{P}} = \langle x, y \mid [x, y] \rangle$. Start with w and use commutator relations to shuffle all z and z^{-1} to the right and left ends of the word, respectively, leaving

$\overline{w} = \overline{w}_0$ in the middle of the word. Each move $\overline{w}_i \mapsto \overline{w}_{i+1}$ that applies the relator $[x, y]$ lifts to a move that introduces a $z^{\pm 1}$ because $[x, y] \in \overline{R}$ lifts to $[x, y]z^{-1} \in R$. Whenever a z or z^{-1} appears in this way, immediately shuffle it to the right or left end of the word, respectively. As they arrive, compress the letters z at the right end of the word as per Lemma 3.1.2, and compress the z^{-1} at the left end by using the $\mathcal{P}$-sequence obtained by inverting every word in the $\mathcal{P}$-sequence of the lemma. Once $(\overline{w}_i)_{i=0}^{\overline{m}}$ has been exhausted we have a compression word at the right end of the word and its inverse at the left end, and the empty word $\overline{w}_{\overline{m}}$ in between. Freely reduce to obtain the empty word.

The total number of $z^{\pm 1}$ being sent to each end in this process is $\preceq n^2$; the lengths of the words stay $\preceq n$ on account of the compression process, Lemma 3.1.2, and the induction hypothesis; and, as required, the total number of application-of-a-relator moves used is $\preceq n^3$, the dominant term in the estimate coming from Lemma 3.1.2. Note that the reason for collecting zs and z^{-1}s at different ends of the word is that the compression process of Lemma 3.1.2 has some steps that are more expensive that others, and having the power of z (or z^{-1}) monotonically increasing avoids crossing these thresholds unduly often.

3.2   Open questions

By Theorem 1.3.4, an affirmative answer to the following question would be a stronger result than Theorem 3.1.1.

Open Question 3.2.1. Do all class c finitely generated nilpotent groups admit (a)synchronous combings with length functions $L(n) \preceq n^c$? (Cf. [61, 89].)

The 3-dimensional integral Heisenberg group cannot have a synchronous combing with length function growing at most linearly fast because that would contradict the n^3 lower bound on its Dehn function — see [5, 47] or Part I of this volume. This makes it a candidate for a natural example of a group exhibiting the behaviour of the example of Bridson discussed in Remark 1.3.6.

Open Question 3.2.2. Is the 3-dimensional integral Heisenberg group synchronously combable? More generally, which nilpotent groups are synchronously combable?

In reference to the second part of 3.2.2 we mention that Holt [41] showed that the finitely generated nilpotent groups that are *asynchronously* combable are those that are virtually abelian. Our next question shows how little we yet know about Dehn functions of nilpotent groups.

Open Question 3.2.3. [6, Question N2] Does there exist a finitely generated nilpotent group whose Dehn function is not $\simeq$-equivalent to n^α for some integer α?

Young [102, Section 5] describes a nilpotent group put forward by Sapir as a candidate to have Dehn function $\simeq n^2 \log n$. The upper bound has been established; the challenge of proving it to be sharp remains.

Open Question 3.2.4. Do linear upper bounds on the filling length functions for nilpotent groups extend to finite presentations of polycyclic groups or, more generally, linear groups?

Progress (in the affirmative direction) towards 3.2.4 would, by Proposition 2.0.2, increase the evidence for the following.

Conjecture 3.2.5. (Gersten [54]) There is a recursive function $f : \mathbb{N} \to \mathbb{N}$ (e.g. $f(n) = 2^n$) such that finite presentations of linear groups all have $\mathrm{Area}(n) \preceq f(n)$.

Chapter 4

Asymptotic Cones

4.1 The definition

In [61] Gromov says of a finitely generated group G with a word metric d:

> "*This space may at first appear boring and uneventful to a geometer's eye since it is discrete and the traditional local (e.g. topological and infinitesimal) machinery does not run in G.*"

Imagine viewing G from increasingly distant vantage points, i.e. scaling the metric by a sequence $\mathbf{s} = (s_i)$ with $s_i \to \infty$. An asymptotic cone is a limit of $\left(G, \frac{1}{s_n} d \right)$, a coalescing of G to a more continuous object that is amenable to attack by "*topological and infinitesimal machinery*", and which "*fills our geometer's heart with joy*" ([61] again).

The asymptotic cones $\mathrm{Cone}_\omega(\mathcal{X}, \mathbf{e}, \mathbf{s})$ of a metric space $(\mathcal{X}, d)$ are defined with reference to three ingredients: the first is a sequence $\mathbf{e} = (e_i)$ of basepoints in $\mathcal{X}$ (if $\mathcal{X}$ is a group, then, by homogeneity, all the e_i may as well be the identity element); the second is a sequence of strictly positive real numbers $\mathbf{s} = (s_i)$ such that $s_i \to \infty$ as $i \to \infty$; and the third is a *non-principal ultrafilter* on $\mathbb{N}$, the magic that forces convergence. Van den Dries and Wilkie [100] recognised the usefulness of non-principal ultrafilters for cutting through delicate arguments concerning extracting a convergent subsequence (with respect to the Gromov-Hausdorff distance) from a sequence of metric spaces in Gromov's proof [58] that groups of polynomial growth are virtually nilpotent.

Definition 4.1.1. A *non-principal ultrafilter* is a finitely additive probability measure on $\mathbb{N}$ that takes values in $\{0, 1\}$ and gives all singleton sets measure 0.

The existence of non-principal ultrafilters is equivalent to the Axiom of Choice (see Exercises 4.1.2) and so we can be barred from describing asymptotic cones explicitly and most proofs involving asymptotic cones are non-constructive. Generally, asymptotic cones are wild beasts: the only finitely generated groups whose asymptotic cones are locally compact metric spaces are virtually nilpotent

groups. [En route to his Polynomial Growth Theorem [58], Gromov proves that
the asymptotic cones of virtually nilpotent groups are proper. Druţu proves the
converse in [37] and she adds [35] that by the Hopf-Rinow Theorem (see, for example, [21, Proposition 3.7]) proper can be replaced by *locally compact* as asymptotic
cones of groups are complete (see [100]) geodesic spaces (see Exercise 4.1.4).] Indeed, some asymptotic cones contain π_1-injective copies of the Hawaiian earring
and (so) have uncountable fundamental groups [24]. But rather than being put
off, one should see these features as part of the spice of the subject.

The way a non-principal ultrafilter $\omega : \mathcal{P}(\mathbb{N}) \to \{0,1\}$ forces convergence is
by *coherently* (see Exercise 4.1.2 (4), below) choosing limit points from sequences
of real numbers: given a sequence (a_i) in $\mathbb{R}$ we say that $a \in \mathbb{R}$ is an ω-*ultralimit*
of (a_i) when $\forall \varepsilon > 0$, $\omega \{i \mid |a - a_i| < \varepsilon\} = 1$, that ∞ is an ω-ultralimit of (a_i)
when $\omega \{i \mid a_i > N\} = 1$ for all $N > 0$, and that $-\infty$ is an ω-ultralimit when
$\omega \{i \mid a_i < -N\} = 1$ for all $N > 0$.

Exercise 4.1.2.

1. Show that if ω is a non-principal ultrafilter on $\mathbb{N}$ and $A, B \subseteq \mathbb{N}$ satisfy
   $\omega(A) = \omega(B) = 1$, then $\omega(A \cap B) = 1$.

2. Establish the existence of non-principal ultrafilters ω on $\mathbb{N}$. (*Hint.* Consider
   the set of functions $\omega : \mathcal{P}(\mathbb{N}) \to \{0,1\}$ such that $\omega^{-1}(1)$ is closed under taking
   intersections and taking supersets, and does not include the empty set, but
   does include all $A \subseteq \mathbb{N}$ for which $\mathbb{N} \smallsetminus A$ is finite. Use Zorn's Lemma.)

3. Let ω be a non-principal ultrafilter. Prove that every ω-ultralimit is also a
   limit point in the usual sense.

4. Show that every sequence (a_n) of real numbers has a unique ω-ultralimit
   in $\mathbb{R} \cup \{\pm\infty\}$ denoted $\lim_\omega a_n$. (*Hint.* First show that if $+\infty$ or $-\infty$ is an
   ultralimit then it is the unique ultralimit. Next assume $\pm\infty$ is not an ultra-
   limit and prove the existence of an ultralimit by suitably adapting a proof
   that every bounded sequence of real numbers has a limit point. Finally prove
   uniqueness.)

5. Let $\lim_\omega a_i$ denote the ω-ultralimit of (a_i). Prove that for all $\lambda, \mu \in \mathbb{R}$ and
   sequences $(a_i), (b_i)$

   $$\lim_\omega (\lambda a_i + \mu b_i) = \lambda \lim_\omega a_i + \mu \lim_\omega b_i,$$

   and $\lim_\omega a_i < \lim_\omega b_i$ if and only if $a_i < b_i$ for all n in a set of ω-measure 1.

6. Suppose $f : \mathbb{R} \to \mathbb{R}$ is continuous. Show that for sequences (a_i) of real
   numbers, $f(\lim_\omega a_i) = \lim_\omega f(a_i)$.

Points in $\mathrm{Cone}_\omega(\mathcal{X}, \mathbf{e}, \mathbf{s})$ are equivalence classes of sequences (x_i) in $\mathcal{X}$ such
that $\lim_\omega d(e_i, x_i)/s_i < \infty$, where two sequences $(x_i), (y_i)$ are equivalent if and only
if $\lim_\omega d(x_i, y_i)/s_i = 0$. The metric on $\mathrm{Cone}_\omega(\mathcal{X}, \mathbf{e}, \mathbf{s})$ is also denoted by d and is
given by $d(\mathbf{x}, \mathbf{y}) = \lim_\omega d(x_i, y_i)/s_i$ where (x_i) and (y_i) are sequences representing
$\mathbf{x}$ and $\mathbf{y}$.

Exercise 4.1.3. Check d is well-defined and is a metric on $\mathrm{Cone}_\omega(\mathcal{X}, \mathbf{e}, \mathbf{s})$.

The next exercise gives some important basic properties of asymptotic cones. In particular, it shows that if a metric space is $(1, \mu)$-quasi-isometric to a geodesic metric space (as is the case for a group with a word metric, for example) then its asymptotic cones are all geodesic metric spaces. [A *geodesic* in a metric space $(\mathcal{X}, d)$ is an isometric embedding $\gamma : I \to \mathcal{X}$, where $I \subseteq \mathbb{R}$ is a closed interval (possibly infinite or bi-infinite). We say that $(\mathcal{X}, d)$ is a *geodesic space* if every two points in $\mathcal{X}$ are connected by a geodesic.]

Exercise 4.1.4.

1. Show that a (λ, μ)-quasi-isometry $\Phi : \mathcal{X} \to \mathcal{Y}$ between metric spaces induces λ-bi-Lipschitz homeomorphisms

$$\mathrm{Cone}_\omega(\mathcal{X}, (e_i), \mathbf{s}) \to \mathrm{Cone}_\omega(\mathcal{Y}, (\Phi(e_i)), \mathbf{s}),$$

   for all $\omega, (e_i)$ and $\mathbf{s}$.

2. Show that if $\mathbf{a} = (a_i)$ and $\mathbf{b} = (b_i)$ represent points in $\mathrm{Cone}_\omega(\mathcal{X}, \mathbf{e}, \mathbf{s})$ and $\gamma_i : [0, 1] \to \mathcal{X}$ are geodesics from a_i to b_i, parametrised proportional to arc length, then $\boldsymbol{\gamma} : [0, 1] \to \mathrm{Cone}_\omega(\mathcal{X}, \mathbf{e}, \mathbf{s})$ defined by $\boldsymbol{\gamma}(t) = (\gamma_i(t))$ is a geodesic from $\mathbf{a}$ to $\mathbf{b}$.

It is natural to ask how asymptotic cones depend on ω and $\mathbf{s}$. Varying one is similar to varying the other (see [92, Appendix B]) but the precise relationship is not clear. In this chapter we will discuss results about statements about all the asymptotic cones of a group for a fixed ω but $\mathbf{s}$ varying. (Other authors fix $\mathbf{s}$ as $(s_i) = (i)$ and vary ω, and others allow both to vary.) This will be crucial because another quirk of the subject is that the topological type of the asymptotic cones of a group, even, may depend on ω or $\mathbf{s}$. Thomas & Velickovic [99] found the first example of a finitely generated group with two non-homeomorphic asymptotic cones — they used an infinite sequence of defining relations satisfying small cancellation to create holes in the Cayley graph on an infinite sequence of scales, and then, depending on whether or not its ultrafilter caused the asymptotic cone to see a similar sequence of scales, it either has non-trivial fundamental group, or is an $\mathbb{R}$-tree. Later Kramer, Shelah, Tent & Thomas [66] found an example of a finitely presented group with the mind-boggling property of having $2^{2^{\aleph_0}}$ non-homeomorphic cones if the Continuum Hypothesis (CH) is false but only one if it is true. Also they showed that under CH a finitely generated group has at most $2^{\aleph_0}$ non-homeomorphic asymptotic cones, a result Druţu & Sapir [38] proved to be sharp when they found a finitely generated group which (independent of CH) has $2^{\aleph_0}$ non-homeomorphic cones. Most recently, Sapir & Ol'shanskii [78, 75] constructed a finitely presented group for which the vanishing of the fundamental groups of their asymptotic cones depends on $\mathbf{s}$ (independent of CH).

4.2   Hyperbolic groups

A metric space $(\mathcal{X}, d)$ is δ-*hyperbolic* in the sense of Gromov [60] if for all $w, x, y, z \in \mathcal{X}$,

$$d(x, w) + d(y, z) \; \leq \; \max\{d(x, y) + d(z, w), d(x, z) + d(y, w)\} + \delta.$$

We say $(\mathcal{X}, d)$ is *hyperbolic* when it is δ-hyperbolic for some $\delta \geq 0$.

This *four-point condition* for hyperbolicity is one of a number of equivalent formulations. A geodesic metric space $(\mathcal{X}, d)$ is *hyperbolic* if and only if it satisfies the *thin-triangles condition*: there exists $\delta > 0$ such that, given any geodesic triangle in $\mathcal{X}$, any one side is contained in a δ-neighbourhood of the other two.

Exercise 4.2.1. Relate the δ that occurs in the thin-triangles condition to the δ in the four-point condition.

We say that a group G with finite generating set X is hyperbolic when $Cay^1(G, X)$ is a hyperbolic metric space or, equivalently, (G, d_X) satisfies the four-point condition for some $\delta \geq 0$.

For a van Kampen diagram Δ over a finite presentation $\mathcal{P}$, define

$$\mathrm{Rad}(\Delta) \; := \; \max\{\, \rho(a, \partial\Delta) \mid \text{vertices } a \text{ of } \Delta \,\},$$

where ρ is the combinatorial metric on $\Delta^{(1)}$. For a null-homotopic word define $\mathrm{Rad}(w)$ be the minimum of $\mathrm{Rad}(\Delta)$ amongst all van Kampen diagrams for w in the usual way, but also define $\overline{\mathrm{Rad}}(w)$ to be the maximum of $\mathrm{Rad}(\Delta)$ amongst all minimal area diagrams for w. Then, as usual, define the corresponding filling functions, the *radius* and *upper radius* functions, $\mathrm{Rad}, \overline{\mathrm{Rad}} : \mathbb{N} \to \mathbb{N}$ for $\mathcal{P}$ to be the maxima of $\mathrm{Rad}(w)$ and $\overline{\mathrm{Rad}}(w)$, respectively, over all null-homotopic words of length at most n. If $\mathrm{Rad}(n) \succeq n$ then $\mathrm{IDiam}(n) \simeq \mathrm{Rad}(n)$, but in hyperbolic groups the extra sensitivity of Rad for low radius diagrams is useful, as we will see in the following theorem, which gives formulations of hyperbolicity for groups in terms of Dehn functions, radius functions, and asymptotic cones. A geodesic metric space $\mathcal{X}$ is an $\mathbb{R}$-*tree* when each pair of points $a, b \in \mathcal{X}$ is joined by a unique geodesic segment $[a, b]$, and if $[a, b]$ and $[b, c]$ are two geodesic segments with unique intersection point b, then $[a, c]$ is the concatenation of $[a, b]$ and $[b, c]$.

Theorem 4.2.2. *Let ω be a non-principal ultra-filter on $\mathbb{N}$. For a group G with finite generating set X, the following are equivalent.*

(1) *G is hyperbolic.*

(2) *For all sequences of real numbers $\mathbf{s} = (s_i)$ with $s_i \to \infty$, the cones $\mathrm{Cone}_\omega(G, \mathbf{1}, \mathbf{s})$ are $\mathbb{R}$-trees.*

(3) *G admits a Dehn presentation (as defined in Section 1.5).*

(4) *There exists $C > 0$ and a finite presentation for G with Dehn function satisfying $\mathrm{Area}(n) \leq Cn$ for all n.*

(5) If $\mathcal{P}$ is a finite presentation for G then the Dehn function of $\mathcal{P}$ satisfies $\mathrm{Area}(n)/n^2 \to 0$ as $n \to \infty$.

(6) There exists $C > 0$ and a finite presentation for G with $\overline{\mathrm{Rad}}(n) \leq C \log n$ for all n.

(7) There exists $C > 0$ and a finite presentation for G with $\mathrm{Rad}(n) \leq C \log n$ for all n.

(8) If $\mathcal{P}$ is a finite presentation for G then the radius function of $\mathcal{P}$ satisfies $\mathrm{Rad}(n)/n \to 0$ as $n \to \infty$.

In fact, all the asymptotic cones of non-virtually-cyclic hyperbolic groups are the same: they are the universal $\mathbb{R}$-trees with $2^{\aleph_0}$ branching at every vertex [39]. The example [99], discussed earlier, shows that quantifying over all $\mathbf{s}$ is necessary in Theorem 4.2.2 2. However, M. Kapovich & Kleiner [65] have shown that if a *finitely presented* group G has one asymptotic cone that is a $\mathbb{R}$-tree then G is hyperbolic.

Proof of Theorem 4.2.2. We begin by proving the equivalence of 1 and 2, which is due to Gromov [60]; subsequent accounts are [33], [36], and [56, Chapter 2, §1]. The proof here is based closely on that of Druţu and, in fact, amounts to not just a group theoretic result but to a characterisation of hyperbolic geodesic metric spaces.

$1 \implies 2$. Assume G is hyperbolic. From Exercise 4.1.4 we know $\mathrm{Cone}_\omega(G, \mathbf{1}, \mathbf{s})$ is a geodesic space. One can show that any four points $\mathbf{w}, \mathbf{x}, \mathbf{y}, \mathbf{z} \in \mathrm{Cone}_\omega(G, \mathbf{1}, \mathbf{s})$ satisfy the four point condition with $\delta = 0$ by applying the four point condition to w_i, x_i, y_i, z_i, where $(w_i), (x_i), (y_i), (z_i)$ are representatives for $\mathbf{w}, \mathbf{x}, \mathbf{y}, \mathbf{z}$, and using properties of $\lim_\omega$ such as those in Exercise 4.1.2 (4). It follows that all geodesic triangles are in $\mathrm{Cone}_\omega(G, \mathbf{1}, s_i)$ are 0-thin and from that one can deduce that $\mathrm{Cone}_\omega(G, \mathbf{1}, s_i)$ is an $\mathbb{R}$-tree.

$2 \implies 1$. Assume $\mathrm{Cone}_\omega(G, \mathbf{1}, \mathbf{s})$ are $\mathbb{R}$-trees for all $\mathbf{s}$. Let d denote the path metric on $Cay^1(G, X)$. Suppose, for a contradiction, there is no $\delta > 0$ such that all the geodesic triangles in $Cay^1(G, X)$ are δ-thin. Then there are geodesic triangles $[x_i, y_i, z_i]$ in $Cay^1(G, X)$ such that, defining s_i to be the infimal distance such that every point on any one side of $[x_i, y_i, z_i]$ is in an s_i-neighbourhood of the other two, we have $s_i \to \infty$.

By compactness and by relabelling if necessary, we can assume that $s_i = d(a_i, b_i)$ for some $a_i \in [x_i, y_i]$ and $b_i \in [y_i, z_i]$. Then

$$t_i := d(a_i, [x_i, z_i]) \geq s_i$$

and $t_i = d(a_i, c_i)$ for some $c_i \in [x_i, z_i]$ — see Figure 4.1. Let $\mathbf{a} = (a_i)$ and $\mathbf{s} = (s_i)$. Define $\mathcal{C} := \mathrm{Cone}_\omega(Cay^1(G, X), \mathbf{a}, \mathbf{s})$, which is an $\mathbb{R}$-tree by hypothesis.

Let γ_i be a geodesic running from y_i to x_i via $a_i = \gamma_i(0)$ at constant speed s_i — that is, $d(a_i, \gamma_i(r)) = |r| s_i$. Define $k_1 := \lim_\omega d(a_i, x_i)/s_i$ and $k_2 :=$

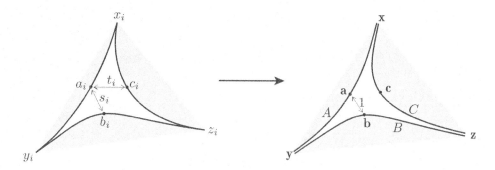

Figure 4.1: The geodesic triangle $[x_i, y_i, z_i]$.

$\lim_\omega d(a_i, y_i)/s_i$, which will be ∞ if $\mathbf{x}$ or $\mathbf{y}$ (respectively) fail to define a point in $\mathcal{C}$. If $k_1, k_2 < \infty$ then let $\boldsymbol{\gamma} : [-k_2, k_1] \to \mathcal{C}$ be the unit speed geodesic defined by

$$\boldsymbol{\gamma}(r) \quad := \quad (\gamma_i(r)) \quad \text{for} \ -k_2 < r < k_1,$$
$$\boldsymbol{\gamma}(k_1) \quad := \quad \mathbf{x}, \quad \boldsymbol{\gamma}(-k_2) := \mathbf{y}.$$

(For $-k_1 < r < k_2$, the expression $\gamma_i(r)$ is well defined for all n in a set of ω-measure 1, and this is enough for $(\gamma_i(r))$ to define a point in $\mathcal{C}$.) If either or both of k_1 and k_2 is infinite, then let $\boldsymbol{\gamma}$ be an infinite or bi-infinite unit-speed geodesic $\boldsymbol{\gamma}(r) = (\gamma_i(r))$. Let A denote the image of $\boldsymbol{\gamma}$. In the same way, define B to be (the image of) the geodesic through $\mathbf{b} = (b_i)$ and between $\mathbf{y}$ and $\mathbf{z}$.

Suppose $\mathbf{c} = (c_i)$ is a well-defined point in $\mathcal{C}$ (that is, $\lim_\omega t_i/s_i < \infty$). Then, as before, define C to be the geodesic through $\mathbf{c}$ and between $\mathbf{x}$ and $\mathbf{z}$. We claim that every point $\mathbf{p} = (p_i)$ on one of A, B, C is in a 1-neighbourhood of the other two. Then, as $\mathcal{C}$ is an $\mathbb{R}$-tree, the only way for A, B and C to remain close in this way, will be for $A \cup B \cup C$ to form a tripod (whose sides may be infinite), but that contradicts $d(\mathbf{a}, B \cup C) = 1$.

Suppose $\mathbf{p}$ is on A. In other words, $\mathbf{p} = \boldsymbol{\gamma}(r) = (\gamma_i(r))$ for some r. We will show $d(p, B \cup C) \le 1$. Similar arguments, which we omit, apply to $\mathbf{p}$ in other locations. Let q_i be a point on $[x_i, z_i] \cup [y_i, z_i]$ closest to p_i. Then $d(p_i, q_i) \le s_i$ and so $d(\mathbf{p}, \mathbf{q}) \le 1$ where $\mathbf{q} = (q_i)$. We will show that $\mathbf{q} \in B \cup C$.

Let $\mathcal{S} = \{i \mid q_i \in [z_i, x_i]\}$ and $\mathcal{S}' = \{i \mid q_i \in [y_i, z_i]\}$. Suppose $\mathcal{S}$ has ω-measure 1. Let $\boldsymbol{\gamma}'$ be the (possibly infinite or bi-infinite) unit speed geodesic in $\mathcal{C}$ with $\boldsymbol{\gamma}'(0) = \mathbf{c}$, running between $\mathbf{z}$ and $\mathbf{x}$, and defined in the same way as $\boldsymbol{\gamma}$. Now $d(p_i, q_i) \le s_i$ and

$$d(c_i, q_i) \quad \le \quad d(c_i, a_i) + d(a_i, p_i) + d(p_i, q_i) \quad \le \quad t_i + |r|\, s_i + s_i.$$

So for all $i \in \mathcal{S}$ we find $q_i = \gamma'(r_i)$ for some r_i with $|r_i| \le t_i/s_i + |r| + 1$ and therefore $|\lim_\omega r_i| < \infty$. It follows that $(\gamma'(r_i)) = (\gamma'(\lim_\omega r_i))$ in $\mathcal{C}$ and hence that $\mathbf{q} \in C$, as required. If $\mathcal{S}$ has ω-measure 0 then $\mathcal{S}'$ has ω-measure 1 and a similar argument shows $\mathbf{q} \in B$.

On the other hand, suppose $\mathbf{c}$ is not a well-defined point in $\mathcal{C}$ or, equivalently, $\lim_\omega t_i/s_i = \infty$. Then A and B are either both infinite or both bi-infinite as $\mathbf{x}$ and $\mathbf{z}$ cannot define points in $\mathcal{C}$. We will show that every point on A is a distance at most 1 from B. In an $\mathbb{R}$-tree this is not compatible with $d(\mathbf{a}, B) = 1$.

Suppose $\mathbf{p} = (p_i)$ is a point on A. Then $\mathbf{p} = \boldsymbol{\gamma}(r) = (\gamma_i(r))$ for some $r \in \mathbb{R}$. As before, let q_i be a point on $[x_i, z_i] \cup [y_i, z_i]$ closest to p_i. For all i such that $q_i \in [c_i, x_i] \cup [c_i, z_i]$ we have $d(a_i, q_i) \geq t_i$. So as $d(\mathbf{p}, \mathbf{q}) = 1$ and $\lim_\omega t_i/s_i = \infty$, there is a set $\mathcal{S} \subseteq \mathbb{N}$ of ω-measure 1 such that $q_i \in [b_i, y_i] \cup [b_i, z_i]$ for all $i \in \mathcal{S}$. A similar argument to that used earlier shows that $\mathbf{q}$ is a point on B, and completes the proof.

$1 \implies 3, 6, 7$. There is an account of an elegant proof due to N. Brady that $1 \implies 3$ in Part I of this volume (Proposition 4.11). That $1 \implies 7$ is proved in Druţu [33] in the course of establishing equivalence of 1, 7 and 8. A proof of $1 \implies 7$ can also be found in Part I of this volume (Lemma 6.2) — the idea is to use the thin-triangles condition to find a finite presentation for G to construct suitable van Kampen diagrams. The stronger result $1 \implies 6$ that the logarithmic radius upper bound is realised on *all* minimal area van Kampen diagrams is stated in [61, §5.C] and proved in [53] using an estimate of the radius of a minimal area diagram in terms of the areas and of the *annular* regions and the lengths of the curves separating them in the decomposition of the proof of Theorem 2.1.13.

$3 \implies 4$, $4 \implies 5$, $6 \implies 7$, and $7 \implies 8$ are all immediate (modulo standard arguments about changing the finite presentation).

$5 \implies 2$. The following argument, combined with $2 \implies 1$, provides an alternative to the better known proofs of $5 \implies 1$, given by Bowditch [14], Ol'shanskii [76] and Papasoglu [84]. The following is a sketch of the proof of Druţu [34] . We begin with two lemmas:

Lemma 4.2.3 (Bowditch [14]). *Suppose we express an edge-circuit γ in the Cayley 2-complex $Cay^2(\mathcal{P})$ of a finite presentation as a union of four consecutive arcs: $\gamma = \alpha^1 \cup \alpha^2 \cup \alpha^3 \cup \alpha^4$. Define $d_1 := d(\alpha^1, \alpha^3)$ and $d_2 := d(\alpha^2, \alpha^4)$ where d is the combinatorial metric on the 1-skeleton of $Cay^2(\mathcal{P})$. Then $\mathrm{Area}(\gamma) \geq K d_1 d_2$, where K is a constant depending only on $\mathcal{P}$.*

Bowditch's lemma can be proved by taking successive star-neighbourhoods S^i of α_1 in a $\mathcal{P}$-van Kampen diagram Δ filling γ. That is, $S^0 = \alpha_1$ and $S^{i+1} = \mathrm{Star}(S^i)$. [For a subcomplex A of a complex B, the star neighbourhood of A is the union of all the (closed) cells that intersect A.] For some constant K', we have $S^i \cap \alpha_3 = \emptyset$ for all $i \leq K' d_1$, and for all such i the portion of ∂S^i in the interior of Δ has length at least d_2. The result then follows by summing estimates for the areas of $S^{i+1} \setminus S^i$.

Our second lemma is straight-forward.

Lemma 4.2.4. *If a geodesic space is not a $\mathbb{R}$-tree, then it contains a geodesic triangle whose sides only meet at their vertices.*

Now assume 5 and suppose $\mathrm{Cone}_\omega(G, \mathbf{1}, \mathbf{s})$ is not an $\mathbb{R}$-tree for some $\mathbf{s}$. So there is a geodesic triangle $[\mathbf{x}, \mathbf{y}, \mathbf{z}]$ satisfying the condition of Lemma 4.2.4. Let l be the sum of the lengths of the sides of $[\mathbf{x}, \mathbf{y}, \mathbf{z}]$. Decompose $[\mathbf{x}, \mathbf{y}, \mathbf{z}]$ into four consecutive arcs $\alpha^1, \alpha^2, \alpha^3, \alpha^4$ such that $d(\alpha^1, \alpha^3), d(\alpha^2, \alpha^4) > \epsilon$ for some $\epsilon > 0$.

Write $\mathbf{x} = (x_i)$, $\mathbf{y} = (y_i)$ and $\mathbf{z} = (z_i)$ where x_i, y_i, z_i are vertices in $Cay^2(\mathcal{P})$. The four points $\mathbf{a}^j = \alpha^j \cap \alpha^{j+1}$ (indices $j = 1, 2, 3, 4$ modulo 4) are represented by sequences $\mathbf{a}^j = (a_i^j)$ of vertices. Connect up $x_i, y_i, z_i, a_i^1, a_i^2, a_i^3, a_i^4$ (in the cyclic order of the corresponding points $\mathbf{x}, \mathbf{y}, \mathbf{z}, \mathbf{a}^1, \mathbf{a}^2, \mathbf{a}^3, \mathbf{a}^4$ around $[\mathbf{x}, \mathbf{y}, \mathbf{z}]$) by geodesics to make an edge-circuit γ_i in $Cay^2(\mathcal{P})$. Let w_i be the word one reads around γ_i and let $l_i := \ell(w_i)$. Then

$$N_1 := \{\, i \mid |l - l_i/s_i| < l/2 \,\}$$

satisfies $\omega(N_1) = 1$. In particular, $s_i > 2l_i/l$ for all $i \in N_1$. Also

$$N_2 := \{\, i \mid d(\alpha_i^1, \alpha_i^3) > \epsilon s_i \,\}$$
$$N_3 := \{\, i \mid d(\alpha_i^2, \alpha_i^4) > \epsilon s_i \,\}$$

have $\omega(N_2) = \omega(N_3) = 1$. So $N := N_1 \cap N_2 \cap N_3$ has measure 1 and so is infinite, and by Lemma 4.2.3 and the inequalities given above,

$$\mathrm{Area}(w_i) \succeq (\epsilon s_i)^2 \geq \epsilon^2 \left(\frac{2l_i}{l} \right)^2 \succeq l_i^2 = \ell(w_i)^2$$

for all $i \in N$. This contradicts 5.

$8 \implies 2$. This is the final step in our proof of the equivalence of the eight statements in the theorem. Assume 8. Let $\mathbf{s} = (s_i)$ be a sequence of real numbers with $s_i \to \infty$. It is straightforward to check that every geodesic triangle $[\mathbf{x}, \mathbf{y}, \mathbf{z}]$ in $\mathrm{Cone}_\omega(G, \mathbf{1}, \mathbf{s})$ is a tripod: lift to a sequence of geodesic triangles $[x_i, y_i, z_i]$ in $Cay^1(G, X)$; each admits a van Kampen diagram with sublinear radius; deduce that each side of $[\mathbf{x}, \mathbf{y}, \mathbf{z}]$ is in a 0-neighbourhood of the other two sides. □

Remarkably, weaker conditions than 5 and 8 imply hyperbolicity. If $\mathcal{P}$ is a finite presentation for which there exist L, N, ϵ such that every null-homotopic word w with

$$N \leq \mathrm{Area}(w) \leq LN$$

has $\mathrm{Area}(w) \leq \epsilon \ell(w)^2$, then $\mathcal{P}$ presents a hyperbolic group [33, 60, 76, 85]. And if there exists $L > 0$ such that $\mathrm{Rad}(n) \leq n/73$ for all $n \leq L$ then $\mathcal{P}$ presents a hyperbolic group [60, 87]; it is anticipated that 8 could replace 73 in this inequality and the result still hold [87].

Exercise 4.2.5. For x, y, z, t in a metric space $(\mathcal{X}, d)$, define $\delta(x, y, z; t)$ to be

$$\max\{\, d(x,t) + d(t,y) - d(x,y),\ d(y,t) + d(t,z) - d(y,z),\ d(z,t) + d(t,x) - d(z,x) \,\}.$$

We say that $(\mathcal{X}, d)$ enjoys the L_δ property of Chatterji [28] if for all $x, y, z \in \mathcal{X}$ there exists $t \in \mathcal{X}$ such that $\delta(x, y, z; t) \leq \delta$.

1. Which of the following spaces are L_δ for some $\delta \geq 0$?

    (a) $\mathbb{Z}^n$ with the word metric associated to the presentation $\langle x_1, \ldots, x_n \mid [x_i, x_j], \forall 1 \leq i < j \leq n \rangle$.

    (b) $\mathbb{R}^n$ with the Euclidean metric.

    (c) A hyperbolic group with a word metric.

2. Prove that if $(\mathcal{X}, d)$ is an L_δ metric space for some $\delta \geq 0$ then all its asymptotic cones are L_0. (In contrast to Theorem 4.2.2, the converse is false for general metric spaces; however it is an open problem for spaces admitting cocompact group actions.)

4.3   Groups with simply connected asymptotic cones

We now give a characterisation of finitely generated groups with simply connected asymptotic cones. The implication *(1)* $\Rightarrow$ *(2)* was proved by R. Handel [63] and by Gromov [61, 5.F]; an account is given by Druţu [34]. The reverse implication appears in [86, page 793]. More general arguments in [92] develop those in [34, 61, 63, 86].

Theorem 4.3.1. *Let G be a group with finite generating set X. Fix any non-principal ultrafilter ω. The following are equivalent.*

1. *The asymptotic cones $\mathrm{Cone}_\omega(G, \mathbf{1}, \mathbf{s})$ are simply connected for all $\mathbf{s} = (s_n)$ with $s_n \to \infty$.*

2. *Let $\lambda \in (0, 1)$. There exist $K, L \in \mathbb{N}$ such that for all null-homotopic words w there is an equality*

$$w = \prod_{i=1}^{K} u_i w_i u_i^{-1} \tag{4.1}$$

    in the free group $F(X)$ for some words u_i and w_i such that the w_i are null-homotopic and $\ell(w_i) \leq \lambda\ell(w) + L$ for all i.

3. *Let $\lambda \in (0, 1)$. There exist $K, L \in \mathbb{N}$ such that for all null-homotopic words w there is a diagram around the boundary of which reads w, and that possesses at most K 2-cells, the boundary circuits of which are labelled by null-homotopic words of length at most $\lambda\ell(w) + L$.*

Sketch proof. The equivalence of *(2)* and *(3)* when the λ's are each the same is proved in the same way as Lemma 1.4.1. We leave the task of extending this to the case where the λ's differ as an exercise.

We sketch a proof by contradiction, essentially from [17], that shows that *(1)* implies *(3)*. Fix $\lambda \in (0, 1)$. Suppose there are null-homotopic words w_n with $\ell(w_n) \to \infty$ and such that if Δ_n is a diagram with $\partial\Delta_n$ labelled by w_n and whose 2-cells have boundaries labelled by null-homotopic words of length at most $\lambda\ell(w)$, then $\mathrm{Area}(\Delta_n) \geq n$. (Note that, despite the lack of mention of L in this last statement, its negation implies (3).) Let $\gamma_n : \partial([0,1]^2) \to Cay^1(G, X)$ be an edge-circuit based at the identity that follows a path labelled by w_n. Define $s_n := \ell(w_n)$ and define $\gamma : \partial([0,1]^2) \to \mathrm{Cone}_\omega(G, \mathbf{1}, \mathbf{s})$ by $\gamma(r) = (\gamma_n(r))$. As $\mathrm{Cone}_\omega(G, \mathbf{1}, \mathbf{s})$ is simply connected, γ can be extended to a continuous map $\overline{\gamma} : [0,1]^2 \to \mathrm{Cone}_\omega(G, \mathbf{1}, \mathbf{s})$. By uniform continuity there exists $\epsilon > 0$ such that for all $a, b \in [0,1]^2$, if $d(a, b) < \epsilon$ (in the Euclidean metric), then $d(\overline{\gamma}(a), \overline{\gamma}(b)) \leq \lambda/16$. For convenience, we can assume $1/\epsilon$ is an integer. Consider the images $\mathbf{x}^{i,j}$ under $\overline{\gamma}$ of the $(1 + 1/\epsilon)^2$ lattice points of the subdivision of $[0,1]^2$ into squares of side $1/\epsilon$. The images $\mathbf{x}^{i,j}, \mathbf{x}^{i+1,j}, \mathbf{x}^{i,j+1}, \mathbf{x}^{i+1,j+1}$ of four corners of a square of side $1/\epsilon$ are represented by sequences $\left(x_n^{i,j}\right), \left(x_n^{i+1,j}\right), \left(x_n^{i,j+1}\right), \left(x_n^{i+1,j+1}\right)$ with

$$\lim_\omega \frac{d\left(x_n^{i,j}, x_n^{i+1,j}\right) + d\left(x_n^{i+1,j}, x_n^{i+1,j+1}\right) + d\left(x_n^{i+1,j+1}, x_n^{i,j+1}\right) + d\left(x_n^{i,j+1}, x_n^{i,j}\right)}{s_n} \leq \frac{\lambda}{4}$$

and hence

$$d\left(x_n^{i,j}, x_n^{i+1,j}\right) + d\left(x_n^{i+1,j}, x_n^{i+1,j+1}\right) + d\left(x_n^{i+1,j+1}, x_n^{i,j+1}\right) + d\left(x_n^{i,j+1}, x_n^{i,j}\right) < \frac{\lambda\ell(w_n)}{2}$$

for all n in a set $\mathcal{S}$ of ω-measure 1 (and hence infinite). Moreover $\mathcal{S}$ can be taken to be independent of the choice of square, as there are only finitely many in the subdivision of $[0,1]^2$. All adjacent $x_n^{i,j}$ can be joined by geodesics to give fillings of infinitely many w_n with diagrams with $1/\epsilon^2$ 2-cells, each with boundary length at most $\lambda\ell(w_n)$. This is the contradiction we seek.

Finally, we assume (3) and sketch a proof of (1). Fix $\lambda \in (0, 1)$ and let K, L be as per (3). Let s_n be a sequence of real numbers with $n \to \infty$. We will show that $\mathcal{C} := \mathrm{Cone}_\omega(G, \mathbf{1}, \mathbf{s})$ is simply connected. We begin by proving that all rectifiable (that is, finite length) loops $\gamma : \partial\mathbb{D}^2 \to \mathcal{C}$ are null-homotopic.

Assume γ is parametrised proportional to arc length. Inscribe a regular m-gon in $\mathbb{D}^2$ with $m > 2/\lambda$. Let $\mathbf{a}^0, \ldots, \mathbf{a}^{m-1}$ be the images under γ of the vertices of the m-gon. For $i = 0, \ldots, m-1$, let $(a_n^i)_{n\in\mathbb{N}}$ be sequences of vertices in $Cay^1(G, X)$ representing $\mathbf{a}^i$. For fixed n and for all i, join a_n^i to a_n^{i+1} (upper indices modulo m) by a geodesic and fill the resulting edge-circuit as in (3). Roughly speaking, as there are no more than K 2-cells in each filling, one topological configuration occurs for all n in some set of ω-measure 1, and the fillings converge in $\mathcal{C}$ to a filling of the geodesic m-gon by a diagram of that type in which the lengths of the boundaries of the $\leq K$ 2-cells are all at most $\lambda\ell(\gamma)$ (the additive L term disappears in the limit). Adding in the regions between the m-gon and γ, we have a filling with at most $K + m$ regions each with boundary length at most $\lambda\ell(\gamma)$. Now iterate the process, refining the filling further, each successive time decreasing the length of the boundaries of the regions by a factor of λ. Asymptotic cones are complete metric spaces [100], and this fact is used in defining a limit that is a continuous extension of γ across $\mathbb{D}^2$.

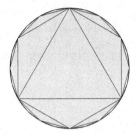

Figure 4.2: A tessellation of $\mathbb{D}^2$ by ideal triangles.

To show that arbitrary loops $\gamma : \partial \mathbb{D}^2 \to \mathcal{C}$ are null-homotopic, first tessellate $\mathbb{D}^2$ by ideal triangles as illustrated in Figure 4.2, and then extend γ to the 1-skeleton of the tessellation by mapping each edge to a geodesic connecting its endpoints. Each geodesic triangle is a rectifiable loop of length at most three times the diameter of the image of γ (which is finite, by compactness). So each triangle can be filled in the way already explained, and the result is an extension of γ to a continuous map $\overline{\gamma} : \mathbb{D}^2 \to \mathcal{C}$. This extension can be proved to be continuous by an argument that uses continuity of γ to estimate the perimeter of geodesic triangles in the tessellation whose domains are close to $\partial \mathbb{D}^2$. Details are in [92]. □

Exercise 4.3.2. Recall that if X and X' are two finite generating sets for G, then (G, d_X) and (G, d'_X) are quasi-isometric and so their asymptotic cones are Lipschitz-equivalent. It follows that conditions *(2)* and *(3)* of Theorem 4.3.1 do not depend on the particular finite generating set X. Prove this result directly, i.e. without using asymptotic cones.

The polynomial bound (4.2) of the following theorem is due to Gromov [61, $5F''_1$]. Druţu [34, Theorem 5.1] also provides a proof. The upper bound on IDiam appears as a remark of Papasoglu at the end of [86] and is proved in detail in [17]. The idea is to repeatedly refine the 2-cells of a diagram as per Theorem 4.3.1 (3) until the lengths of the boundary loops of the 2-cells are all at most a constant, but there is a technical issue of maintaining planarity during each refinement. Below, we use the Dehn proof system to avoid this issue (or, to be more honest, push it elsewhere: to the proof of Proposition 1.5.2, essentially).

Theorem 4.3.3. *Fix any non-principal ultrafilter ω. Suppose, for all sequences of scalars $\mathbf{s}$ with $s_n \to \infty$, the asymptotic cones $\mathrm{Cone}_\omega(G, \mathbf{1}, \mathbf{s})$ of a finitely generated group G are simply connected. Then there exists a finite presentation $\langle X \mid R \rangle$ for G with respect to which*

$$\mathrm{Area}(n) \preceq n^\alpha, \tag{4.2}$$
$$\mathrm{IDiam}(n) \preceq n, \tag{4.3}$$
$$\mathrm{FL}(n) \preceq n, \tag{4.4}$$

where $\alpha = \log_{1/\lambda} K$ and K is as in Theorem 4.3.1. Further, these bounds are realisable simultaneously.

Proof. Fix $\lambda \in (0,1)$ and let $K, L \in \mathbb{N}$ be as in Theorem 4.3.1 (3). Suppose w is null-homotopic. Then, for similar reasons to those used in Section 1.5, there is a sequence $(w_i)_{i=0}^m$ with each w_{i+1} obtained from w_i by either a free reduction/expansion or by

$$w_i = \alpha\beta \mapsto \alpha u\beta = w_{i+1} \tag{4.5}$$

where u is a null-homotopic word with $\ell(u) \le \lambda\ell(w) + L$. Moreover, the number of moves of type (4.5) is at most K and

$$\max_i \ell(w_i) \le (K+1)(\lambda n + L) + n.$$

This bound on $\max_i \ell(w_i)$ holds because the total number of edges in the diagram is at most $K(\lambda n + L) + n$, and a further $\lambda n + L$ is added because the moves of type (4.5) insert the whole of u rather than exchanging one part of a relator for another like an application-of-a-relator move.

Let R be the set of null-homotopic words on X of length at most $1 + (L/(1-\lambda))$. We will show by induction on $\ell(w)$ that w admits a null-$\mathcal{P}$-sequence where $\mathcal{P} = \langle X \mid R \rangle$. This is clearly the case if $\ell(w) \le 1 + (L/(1-\lambda))$. Assume $\ell(w) > 1 + (L/(1-\lambda))$. Then for u as in (4.5)

$$\ell(u) \le \lambda\ell(w) + L < \ell(w),$$

and so by the induction hypothesis and Proposition 1.5.2, each u admits a null-sequence involving at most $\mathrm{Area}(u)$ application-of-a-relator moves and words of length at most $FL(u)$. In place of each move (4.5) insert such a null-sequence, run backwards, to create the u. The result is a null-$\mathcal{P}$-sequence for w. Moreover, this null-$\mathcal{P}$-sequence shows that

$$\mathrm{Area}(n) \le K\,\mathrm{Area}(\lambda n + L) \quad \text{and} \tag{4.6}$$
$$FL(n) \le FL(\lambda n + L) + (K+1)(\lambda n + L) + n. \tag{4.7}$$

Let $k = 1 + \lfloor \log_\lambda(1/n) \rfloor$. Repeatedly apply (4.6) until, after k iterations, we have $\mathrm{Area}(n)$ in terms of

$$\mathrm{Area}\left(\lambda^k n + \frac{L}{1-\lambda}\right).$$

But this is 1 (or 0) because the argument is at most $1 + (L/(1-\lambda))$. So

$$\mathrm{Area}(n) \le K^k \le K^{1-\log_\lambda n} = K\, n^{\log_{1/\lambda} K},$$

which proves (4.2). We get (4.4) by summing a geometric series arising from iteratively applying (4.7), and (4.3) then follows. $\square$

Example 4.3.4. Finitely generated groups (G, X) satisfying the L_δ condition of Exercise 4.2.5 enjoy conditions (*2*) and (*3*) of Theorem 4.3.1 with $K = 3$, with $L = \delta$ and $\lambda = 2/3$. So all the asymptotic cones of G are simply connected, G is finitely presentable, and

$$\text{Area}(n) \preceq n^{\log_{3/2} 3}, \quad \text{FL}(n) \preceq n, \quad \text{and} \quad \text{IDiam}(n) \preceq n.$$

This upper bound on Dehn function is subcubic ($\log_{3/2} 3 \simeq 2.71$) and is due to Elder [40]. It is an open problem to find a group satisfying the L_δ condition for some δ but not having $\text{Area}(n) \preceq n^2$.

Having Dehn function growing at most polynomially fast does not guarantee that the asymptotic cones of a group G will all be simply connected: there are groups [17, 95] with such Dehn functions but with IDiam growing faster than linearly. Indeed, Ol'shanskii and Sapir have recently constructed a group with with $\text{IDiam}(n) \simeq n$ and $\text{Area}(n) \simeq n^3$ but no asymptotic cones simply connected [79]:

$$\langle\, a, b, c, k \;\mid\; [a, b],\, [a, c],\, k^b = ka,\, k^c = ka \,\rangle.$$

They claim that S-machines from [80] could be be used to obtain an example with $\text{IDiam}(n) \simeq n$ and $\text{Area}(n) \simeq n^2 \log n$. So the following theorem of Papasoglu [86] is near sharp.

Theorem 4.3.5. *If a group has quadratic Dehn function, then its asymptotic cones are all simply connected.*

Papasoglu's proof shows that groups with quadratic Dehn function have linear IDiam and it then proceeds to the criterion in Theorem 4.3.1. In particular, by Theorem 4.3.3, one can deduce (without using asymptotic cones) that groups with quadratic Dehn functions have linear FL, and this seems a non-trivial result. More generally, it is shown in [52] that groups with $\text{Area}(n) \preceq n^\alpha$ for some $\alpha \geq 2$ have $\text{IDiam}(n) \preceq n^{\alpha-1}$, and in [54, Theorem 8.2] the additional conclusion $\text{GL}(n) \preceq n^{\alpha-1}$ is drawn.

Exercise 4.3.6. Give a direct proof that finite presentations with $\text{Area}(n) \preceq n^2$ have $\text{IDiam}(n) \preceq n$, and (*harder*) $\text{FL}(n) \preceq n$.

4.4   Higher dimensions

The notions of finite generability and finite presentability are $\mathcal{F}_1$ and $\mathcal{F}_2$ in a family of *finiteness conditions* $\mathcal{F}_n$ for a group G. We say G is of *type* $\mathcal{F}_n$ if there is a $K(G, 1)$ which is a complex with finite n-skeleton. (A $K(G, 1)$ is a space with fundamental group G and all other homotopy groups trivial.) For example, if G has finitely presentation $\mathcal{P}$ then one can attach 3-cells, then 4-cells, and so on, to the presentation 2-complex of $\mathcal{P}$ in such a way as to kill off π_2, then π_3, and so on, of its universal cover, producing a $K(G, 1)$ with finite 1- and 2-skeleta.

Suppose a group G is of type $\mathcal{F}_n$ for some $n \geq 2$. Then there are reasonable notions of higher dimensional filling inequalities concerning filling $(n-1)$-spheres with n-discs in some appropriate sense. The following approach of Bridson [19] is a geometric/combinatorial interpretation of the algebraic definitions of Alonso, Pride & Wang [3]. By way of warning we remark that there is no clear agreement between this definition and that of Epstein et al. [41], that concerns spheres in a Riemannian manifold on which G acts properly, discontinuously and cocompactly.

Starting with a finite presentation for G build a *presentation $(n+1)$-complex* by attaching finitely many 3-cells to kill π_2 of the presentation 2-complex, and then finitely many 4-cells to kill π_3, and so on up to $(n+1)$-cells to kill π_n. The data involved in this construction is termed a *finite $(n+1)$-presentation $\mathcal{P}_{n+1}$*. Let $Cay^{n+1}(\mathcal{P}_{n+1})$ denote the universal cover of the presentation $(n+1)$-complex of $\mathcal{P}_{n+1}$. An important technicality here is that the $(j+1)$-cells may be attached not by combinatorial maps from their boundary combinatorial n-spheres, but by *singular combinatorial maps*. These can collapse i-cells, mapping them into the $(i-1)$-skeleton of their range — see [19] for more details.

Consider a singular combinatorial map $\gamma : S^n \to Cay^{n+1}(\mathcal{P}_{n+1})$, where S^n is some combinatorial n-sphere. We can *fill* γ by giving a singular combinatorial extension $\bar{\gamma} : D^{n+1} \to Cay^{n+1}(\mathcal{P}_{n+1})$ with respect to some combinatorial $(n+1)$-disc D^{n+1} such that $S^n = \partial D^{n+1}$. We define $\mathrm{Vol}_n(\gamma)$ and $\mathrm{Vol}_{n+1}(\bar{\gamma})$ to be the number of n-cells e in S^n and D^{n+1}, respectively, such that $\gamma|_e$ is a homeomorphism, and we define the *filling volume* $\mathrm{FVol}(\gamma)$ of γ to be the minimum amongst all $\mathrm{Vol}_{n+1}(\bar{\gamma})$ such that $\bar{\gamma}$ fills γ. This leads to the definition of the n-th order Dehn function, which, up to $\simeq$ equivalence, is independent of the choice of finite $(n+1)$-presentation [3].

We will be concerned with inequalities concerning not just filling volume but also (intrinsic) diameter. So define $\mathrm{Diam}_n(\gamma)$ and $\mathrm{Diam}_{n+1}(\bar{\gamma})$ to be the maxima of the distances between two vertices on S^n or D^{n+1}, respectively, in the combinatorial metric on their 1-skeleta. And define $\mathrm{FDiam}(\gamma)$ to be the minimum of $\mathrm{Diam}_{n+1}(\bar{\gamma})$ amongst all $\bar{\gamma}$ filling γ.

Recall from Section 1.3 how we coned off a loop (a 1-sphere) as in Figure 1.8 to get an upper bound on the Dehn function. In the same way it is possible to cone off a singular combinatorial n-sphere and then fill the *rods* (the cones over each of the $\mathrm{Vol}_n(\gamma)$ non-collapsing n-cells) to prove the following generalisation of Theorem 1.3.4.

Theorem 4.4.1. [41, 48] *Suppose G is a finitely generated, asynchronously combable group with length function $k \mapsto \mathrm{L}(k)$. Then G is of type $\mathcal{F}_n$ for all n. Further, given a finite $(n+1)$-presentation $\mathcal{P}_{n+1}$ for G, every singular combinatorial n-sphere $\gamma : S^n \to Cay^{n+1}(\mathcal{P}_{n+1})$ can be filled by some singular combinatorial $(n+1)$-disc $\bar{\gamma} : D^{n+1} \to Cay^{n+1}(\mathcal{P}_{n+1})$ with*

$$\mathrm{FVol}_{n+1}(\gamma) \preceq \mathrm{Vol}_n(\gamma)\,\mathrm{L}(\mathrm{Diam}_n(\gamma)),$$
$$\mathrm{FDiam}_{n+1}(\gamma) \preceq \mathrm{L}(\mathrm{Diam}_n(\gamma)).$$

Moreover, these bounds are realisable simultaneously.

Similarly, by coning off and then filling the *rods* it is possible to obtain filling inequalities for groups with n-connected asymptotic cones. (A space is *n-connected* if its homotopy groups $\pi_0, \pi_1, \ldots, \pi_n$ are all trivial.) Each rod has filling volume that is at most polynomial in its diameter for reasons similar to those that explain the appearance of the polynomial area bound in Theorem 4.3.3.

Theorem 4.4.2. [92] *Suppose G is a finitely generated group whose asymptotic cones are all n-connected ($n \geq 1$). Then G is of type $\mathcal{F}_{n+1}$ and, given any finite $(n+1)$-presentation $\mathcal{P}_{n+1}$ for G, every singular combinatorial n-sphere $\gamma : S^n \to Cay^{n+1}(\mathcal{P}_{n+1})$ can be filled by some singular combinatorial $(n+1)$-disc $\bar{\gamma} : D^{n+1} \to Cay^{n+1}(\mathcal{P}_{n+1})$ with*

$$\mathrm{FVol}_{n+1}(\gamma) \preceq \mathrm{Vol}_n(\gamma)\,(\mathrm{Diam}_n(\gamma))^{\alpha_n},$$
$$\mathrm{FDiam}_{n+1}(\gamma) \preceq \mathrm{Diam}_n(\gamma),$$

for some α_n depending only on $\mathcal{P}_{n+1}$. Moreover, these bounds are realisable simultaneously.

Open Question 4.4.3. Do the higher order Dehn functions of hyperbolic groups admit linear upper bounds?

Perhaps the characterisation of hyperbolic groups in terms of $\mathbb{R}$-trees in Theorem 4.2.2 can be used to resolve this question. Mineyev [70] gets linear upper bounds on the volumes of *homological* fillings.

Open Question 4.4.4. Is there a sequence G_n of groups (discrete if possible) such that the asymptotic cones of G_n are all n-connected but not all $(n+1)$-connected?

Gromov [61, §2.B_1] makes the tantalising suggestion that $\mathrm{SL}_n(\mathbb{Z})$ might provide a family of examples. Epstein & Thurston [41, Chapter 10] show, roughly speaking, that any $(n-2)$-st order isoperimetric function for $\mathrm{SL}_n(\mathbb{Z})$ is at least exponential and hence, by Theorem 4.4.2, the asymptotic cones of $\mathrm{SL}_n(\mathbb{Z})$ are not all $(n-2)$-connected. It remains to show that all the asymptotic cones of $\mathrm{SL}_n(\mathbb{Z})$ are $(n-3)$-connected.

In [61, §2.B.(f)] Gromov outlines a strategy for showing that a certain sequence of solvable Lie groups S_n will have every asymptotic cone $(n-3)$-connected but of non-trivial (uncountably generated, even) π_{n-2}. The reason one expects the non-triviality of π_{n-2} is that similar arguments to those in [41] should show the $(n-2)$-nd order isoperimetric function of S_n again to be at least exponential.

Stallings gave the first example of a finitely presented group that is not of type F_3 [98]. Bieri generalised this to a family of groups SB_n of type F_{n-1} but not F_n [9]. These groups are discussed in Part III of this volume. Their asymptotic cones are ripe for investigation.

Bibliography

[1] D. Allcock. An isoperimetric inequality for the Heisenberg groups. *Geom. Funct. Anal.* **8** (1998), no. 2, 219–233.

[2] J.M. Alonso. Inégalitiés isopérimétriques et quasi-isométries. *C.R. Acad. Sci. Paris Ser. 1 Math.* **311** (1990), 761–764.

[3] J.M. Alonso, S.J. Pride, and X. Wang. Higher-dimensional isoperimetric (or Dehn) functions of groups. *J. Group Theory* **2** (1999), 81–122.

[4] G. Baumslag. A non-cyclic one-relator group all of whose finite quotients are cyclic. *J. Austral. Math. Soc.* **10** (1969), 497–498.

[5] G. Baumslag, C.F. Miller, III, and H. Short. Isoperimetric inequalities and the homology of groups. *Invent. Math.* **113** (1993), no. 3, 531–560.

[6] G. Baumslag, A.G. Myasnikov, and V. Shpilrain. Open problems in combinatorial group theory. Second edition. *Contemp. Math., Amer. Math. Soc.* **296** (2002), 1–38. http://algebraweb.info/.

[7] G. Baumslag and D. Solitar. Some two-generator one-relator non-Hopfian groups. *Bull. Amer. Math. Soc.* **68** (1962), 199–201.

[8] A.A. Bernasconi. *On HNN-extensions and the complexity of the word problem for one-relator groups.* PhD thesis, University of Utah, 1994. http://www.math.utah.edu/~sg/Papers/bernasconi-thesis.pdf.

[9] R. Bieri. *Homological dimension of discrete groups.* Queen Mary Lecture Notes, 1976.

[10] J.-C. Birget. Time-complexity of the word problem for semigroups and the Higman embedding theorem. *Internat. J. Algebra Comput.* **8** (1998), no. 2, 235–294.

[11] J.-C. Birget. Functions on groups and computational complexity. *Internat. J. Algebra Comput.* **14** (2004), no. 4, 409–429.

[12] J.-C. Birget, A. Yu. Ol'shanskii, E. Rips, and M. V. Sapir. Isoperimetric functions of groups and computational complexity of the word problem. *Ann. of Math. (2)* **156** (2002), no. 2, 467–518.

[13] W.W. Boone. Certain simple unsolvable problems in group theory I, II, III, IV, V, VI. *Nederl. Akad. Wetensch Proc. Ser. A.* **57** (1954), 231–236, 492–497, **58** (1955), 252–256, 571–577, **60** (1957), 22–26, 227–232.

[14] B.H. Bowditch. A short proof that a subquadratic isoperimetric inequality implies a linear one. *Michigan Math. J.* **42** (1995), no. 1, 103–107.

[15] M.R. Bridson. On the geometry of normal forms in discrete groups. *Proc. London Math. Soc.* **67** (1993), no. 3, 596–616.

[16] M.R. Bridson. *Area versus diameter for van Kampen diagrams.* Unpublished Notes, Oxford University, 1997.

[17] M.R. Bridson. Asymptotic cones and polynomial isoperimetric inequalities. *Topology* **38** (1999), no. 3, 543–554.

[18] M.R. Bridson. The geometry of the word problem. In *Invitations to Geometry and Topology*, ed. by M.R. Bridson and S.M. Salamon, pp. 33–94. Oxford Univ. Press, 2002.

[19] M.R. Bridson. Polynomial Dehn functions and the length of asynchronously automatic structures. *Proc. London Math. Soc.* **85** (2002), no. 2, 441–465.

[20] M.R. Bridson. Combings of groups and the grammer of reperameterization. *Comment. Math. Helv.* **78** (2003), no. 4, 752–771.

[21] M.R. Bridson and A. Haefliger. *Metric Spaces of Non-positive Curvature.* Grundlehren der mathematischen Wissenschaften 319, Springer-Verlag, 1999.

[22] M.R. Bridson and T.R. Riley. Extrinsic versus intrinsic diameter for Riemannian filling-discs and van Kampen diagrams. Preprint, `arXiv:math.GR/0511004`.

[23] M.R. Bridson and T.R. Riley. Free and fragmenting filling length. *Journal of Algebra.* To appear. `arXiv:math.GR/0512162`.

[24] J. Burillo. *Dimension and Fundamental Groups of Asymptotic Cones.* PhD thesis, University of Utah, 1996.

[25] J. Burillo and J. Taback. Equivalence of geometric and combinatorial Dehn functions. *New York Journal of Mathematics* **8** (2002), 169–179.

[26] A. Carbone. Geometry and combinatorics of proof structures. In preparation.

[27] A. Carbone. Group cancellation and resolution. *Studia Logica* **82** (2006), no. 1, 73–93.

[28] I. Chatterji. *On Property (RD) for certain discrete groups.* PhD thesis, L'Université de Lausanne, 2001.

[29] D.E. Cohen. Isoperimetric and isodiametric inequalities for group presentations. *Int. J. of Alg. and Comp.* **1** (1991), no. 3, 315–320.

[30] D.E. Cohen, K. Madlener, and F. Otto. Separating the intrinsic complexity and the derivational complexity of the word problem for finitely presented groups. *Math. Logic Quart.* **39** (1993), no. 2, 143–157.

[31] G. Conner. Isoperimetric functions for central extensions. In *Geometric Group Theory*, ed. by R. Charney, M. Davis, and M. Shapiro. *Ohio State University, Math. Res. Inst. Publ.* 3, pp. 73–77. De Gruyter, 1995.

[32] M. Dehn. Über unendliche diskontunuierliche Gruppen. *Math. Ann.* **71** (1912), 116–144.

[33] C. Druţu. Cône asymptotique et invariants de quasi-isométrie. Preprint, Université de Lille I, 1999.

[34] C. Druţu. Cônes asymptotiques et invariants de quasi-isométrie pour des espaces métriques hyperboliques. *Ann. Inst. Fourier Grenoble* **51** (2001), 81–97.

[35] C. Druţu. Personal communication.

[36] C. Druţu. *Réseaux dans groupes de Lie semisimples et invariants de quasi-isométrie.* PhD thesis, Université de Paris-Sud XI.

[37] C. Druţu. Quasi-isometry invariants and asymptotic cones. *Internat. J. Algebra Comput.* **12** (2002), no. 1–2, 99–135. International Conference on Geometric and Combinatorial Methods in Group Theory and Semigroup Theory (Lincoln, NE, 2000).

[38] C. Druţu and M. Sapir. Tree-graded spaces and asymptotic cones of groups. *Topology* **44** (2005), no. 5, 959–1058. With an appendix by D. Osin and M. Sapir.

[39] A.G. Dyubina and I.V. Polterovich. Structures at infinity of hyperbolic spaces. *Uspekhi Mat. Nauk.* **53** (1998), no. (5, 323), 239–240.

[40] M. Elder. L_δ groups are almost convex and have a sub-cubic Dehn function. *Algebr. Geom. Topol.* **4** (2004), 23–29 (electronic).

[41] D.B.A. Epstein, J.W. Cannon, D.F. Holt, S.V.F. Levy, M.S. Paterson, and W.P. Thurston. *Word Processing in Groups.* Jones and Bartlett, 1992.

[42] B. Farb. Automatic groups: a guided tour. *Enseign. Math. (2)* **38** (1992), no. 3–4, 291–313.

[43] S. Frankel and M. Katz. The Morse landscape of a Riemannian disc. *Ann. Inst. Fourier, Grenoble* **43** (1993), no. 2, 503–507.

[44] S.M. Gersten. The double exponential theorem for isoperimetric and isodiametric functions. *Int. J. of Alg. and Comp.* **1** (1991), no. 3, 321–327.

[45] S.M. Gersten. Isodiametric and isoperimetric inequalities in group extensions. Preprint, University of Utah, 1991.

[46] S.M. Gersten. Dehn functions and l_1-norms of finite presentations. In *Algorithms and classification in combinatorial group theory (Berkeley, CA, 1989)*, ed. by G. Baumslag and C.Miller. *Math. Sci. Res. Inst. Publ.* 3, pp. 195–224. Springer-Verlag, 1992.

[47] S.M. Gersten. Isoperimetric and isodiametric functions. In *Geometric group theory I*, ed. by G. Niblo and M. Roller. LMS lecture notes 181. Cambridge Univ. Press, 1993.

[48] S.M. Gersten. Finiteness properties of asynchronously automatic groups. In *Geometric Group Theory*, ed. by R. Charney, M. Davis, and M. Shapiro. *Ohio State University, Mathematical Research Institute Publications* 3, pp. 121–133. De Gruyter, 1995.

[49] S.M. Gersten. Introduction to hyperbolic and automatic groups. In *Summer School in Group Theory in Banff, 1996. CRM Proc. Lecture Notes* 17, pp. 45–70. Amer. Math. Soc., Providence, RI, 1999.

[50] S.M. Gersten, D.F. Holt, and T.R. Riley. Isoperimetric functions for nilpotent groups. *GAFA* **13** (2003), 795–814.

[51] S.M. Gersten and T.R. Riley. The gallery length filling function and a geometric inequality for filling length. *Proc. London Math. Soc.* **93** (2006), no. 3, 601–623.

[52] S.M. Gersten and T.R. Riley. Filling length in finitely presentable groups. *Geom. Dedicata* **92** (2002), 41–58.

[53] S.M. Gersten and T.R. Riley. Filling radii of finitely presented groups. *Quart. J. Math. Oxford* **53** (2002), no. 1, 31–45.

[54] S.M. Gersten and T.R. Riley. Some duality conjectures for finite graphs and their group theoretic consequences. *Proc. Edin. Math. Soc.* **48** (2005), no. 2, 389–421.

[55] S.M. Gersten and H. Short. Some isoperimetric inequalities for free extensions. *Geom. Dedicata* **92** (2002), 63–72.

[56] E. Ghys and P. de la Harpe. *Sur les groupes hyperboliques d'après M. Gromov.* Birkhäuser, 1990.

[57] R.H. Gilman. Formal languages and their application to combinatorial group theory. In *Groups, languages, algorithms. Contemp. Math.* 378, pp. 1–36. Amer. Math. Soc., Providence, RI, 2005.

[58] M. Gromov. Groups of polynomial growth and expanding maps. *Inst. Hautes Études Sci. Publ. Math.* **53** (1981), 53–78.

[59] M. Gromov. Filling Riemannian manifolds. *J. Differential Geom.* **18** (1983), no. 1, 1–147.

[60] M. Gromov. Hyperbolic groups. In *Essays in group theory*, ed. by S.M. Gersten. *MSRI publications* 8, pp. 75–263. Springer-Verlag, 1987.

[61] M. Gromov. Asymptotic invariants of infinite groups. In *Geometric group theory II*, ed. by G. Niblo and M. Roller. LMS lecture notes 182. Cambridge Univ. Press, 1993.

[62] M. Gromov. *Carnot-Carathéodory spaces seen from within.* Progress in Mathematics 144, pp. 79–323. Birkhäuser, 1996.

[63] R. Handel. Investigations into metric spaces obtained from finitely generated groups by methods of non-standard analysis. M.Sc. Thesis under A. Wilkie, Oxford University.

[64] C. Hidber. Isoperimetric functions of finitely generated nilpotent groups. *J. Pure Appl. Algebra* **144** (1999), no. 3, 229–242.

[65] M. Kapovich and B. Kleiner. Geometry of quasi-planes. Preprint.

[66] L. Kramer, S. Shelah, K. Tent, and S. Thomas. Asymptotic cones of finitely presented groups. *Adv. Math.* **193** (2005), no. 1, 142–173.

[67] R.C. Lyndon and P.E. Schupp. *Combinatorial Group Theory.* Springer-Verlag, 1977.

[68] K. Madlener and F. Otto. Pseudonatural algorithms for the word problem for finitely presented monoids and groups. *J. Symbolic Comput.* **1** (1985), no. 4, 383–418.

[69] A.G. Miasnikov, A. Ushakov, and D. Wong. In preparation. 2005.

[70] I. Mineyev. Higher dimensional isoperimetric functions in hyperbolic groups. *Math. Z.* **233** (2000), no. 2, 327–345.

[71] C.B. Morrey. The problem of Plateau in a Riemann manifold. *Ann. of Math.* **49** (1948), 807–851.

[72] L. Mosher. Mapping class groups are automatic. *Ann. of Math. (2)* **142** (1995), no. 2, 303–384.

[73] P.S. Novikov. On the algorithmic unsolvability of the word problem in group theory. *Trudt Mat. Inst. Stkelov* **44** (1955), 1–143.

[74] K. Ohshika. *Discrete groups.* Translations of Mathematical Monographs 207. American Mathematical Society, Providence, RI, 2002. Translated from the 1998 Japanese original by the author, Iwanami Series in Modern Mathematics.

[75] A.Yu. Ol'shanskii. Groups with quadratic-non-quadratic Dehn functions. Preprint, arXiv:math.GR/0504349.

[76] A.Yu. Ol'shanskii. Hyperbolicity of groups with subquadratic isoperimetric inequality. *Internat. J. Algebra Comput.* **1** (1991), no. 3, 281–289.

[77] A.Yu. Ol'shanskii. On the distortion of subgroups of finitely presented groups. *Mat. Sb.* **188** (1997), no. 11, 51–98.

[78] A.Yu. Ol'shanskii and M.V. Sapir. A finitely presented group with two non-homeomorphic asymptotic cones. Preprint, `arXiv:math.GR/0504350`.

[79] A.Yu. Ol'shanskii and M.V. Sapir. Groups with non-simply connected asymptotic cones. *Contemp. Math., Amer. Math. Soc.* **394** (2006), 203–208.

[80] A.Yu. Ol'shanskii and M.V. Sapir. Groups with small Dehn functions and bipartite chord diagrams. GAFA. To appear, `arXiv:math.GR/0411174`.

[81] A.Yu. Ol'shanskii and M.V. Sapir. Quadratic isometric functions of the Heisenberg groups. A combinatorial proof. *J. Math. Sci. (New York)* **93** (1999), no. 6, 921–927. Algebra, 11.

[82] A.Yu. Ol'shanskii and M.V. Sapir. Length and area functions on groups and quasi-isometric Higman embeddings. *Internat. J. Algebra Comput.* **11** (2001), no. 2, 137–170.

[83] P. Pansu. Croissance des boules et des géodesiques fermées dans les nil-variétés. *Ergodic Theory Dynam. Systems* **3** (1983), 415–445.

[84] P. Papasoglu. On the sub-quadratic isoperimetric inequality. In *Geometric group theory (Columbus, OH, 1992)*. Ohio State Univ. Math. Res. Inst. Publ. 3, pp. 149–157. De Gruyter, Berlin, 1995.

[85] P. Papasoglu. An algorithm detecting hyperbolicity. In *Geometric and computational perspectives on infinite groups (Minneapolis, MN and New Brunswick, NJ, 1994)*. DIMACS Ser. Discrete Math. Theoret. Comput. Sci. 25, pp. 193–200. Amer. Math. Soc., 1996.

[86] P. Papasoglu. On the asymptotic invariants of groups satisfying a quadratic isoperimetric inequality. *J. Differential Geom.* **44** (1996), 789–806.

[87] P. Papasoglu. Quasi-flats in semihyperbolic groups. *Proc. Amer. Math. Soc.* **126** (1998), no. 5, 1267–1273.

[88] P. Papasoglu. Isodiametric and isoperimetric inequalities for complexes and groups. *J. London Math. Soc. (2)* **63** (2000), no. 1, 97–106.

[89] Ch. Pittet. Isoperimetric inequalities for homogeneous nilpotent groups. In *Geometric Group Theory*, ed. by R. Charney, M. Davis, and M. Shapiro. *Ohio State University, Mathematical Research Institute Publications* 3, pp. 159–164. De Gruyter, 1995.

[90] A.N. Platonov. An isoperimetric function of the Baumslag-Gersten group. *Vestnik Moskov. Univ. Ser. I Mat. Mekh.* **3** (2004), 12–17, 70. Translation in *Moscow Univ. Math. Bull.* **59** (2004).

[91] T.R. Riley. *Asymptotic invariants of infinite discrete groups.* PhD thesis, Oxford University, 2002.

[92] T.R. Riley. Higher connectedness of asymptotic cones. *Topology* **42** (2003), 1289–1352.

[93] T.R. Riley and W.P. Thurston. The absence of efficient dual pairs of spanning trees in planar graphs. *Electronic J. Comp.* **13**, N13 (2006).

[94] M. Sapir. Algorithmic and asymptotic properties of groups. *Proceedings of the ICM Madrid 2006*, E.M.S. Ph., 2006.

[95] M.V. Sapir, J.-C. Birget, and E. Rips. Isoperimetric and isodiametric functions of groups. *Ann. of Math. (2)* **156** (2002), no. 2, 345–466.

[96] P. Schupp. Personal communication.

[97] H. Short *et al.*. Notes on word hyperbolic groups. In *Group Theory from a Geometrical Viewpoint (Trieste, 1990)*, ed. by E. Ghys, A. Haefliger, and A. Verjovsky. *MSRI publications* 8, pp. 3–63. World Scientific Publishing, River Edge, N.J., 1991.

[98] J. Stallings. A finitely presented group whose 3-dimensional integral homology is not finitely generated. *Amer. J. Math.* **85** (1963), 541–543.

[99] S. Thomas and B. Velickovic. Asymptotic cones of finitely generated groups. *Bull. London Math. Soc.* **32** (2000), no. 2, 203–208.

[100] L. van den Dries and A.J. Wilkie. On Gromov's theorem concerning groups of polynomial growth and elementary logic. *J. Algebra* **89** (1984), 349–374.

[101] E.R. van Kampen. On some lemmas in the theory of groups. *Amer. J. Math.* **55** (1933), 268–273.

[102] R. Young. Scaled relators and Dehn functions for nilpotent groups. Preprint, arXiv:math.GR/0601297.

Part III

Diagrams and Groups

Hamish Short

Introduction

These notes are intended to be an introduction to the use of diagrams to study problems in the domain of finitely presented groups, in particular decision problems such as the word problem. The main topics are: van Kampen diagrams, isoperimetric inequalities, small cancellation theory and hyperbolic groups. To illustrate the ideas some examples will be studied in detail: a polynomial lower bound for the isoperimetric inequality for free nilpotent groups (studied by Tim Riley in his part), and a polynomial upper bound for the isoperimetric inequality for certain normal subgroups of hyperbolic groups. Details concering some of the topics touched on during the course are given here : I would like to thank Tim Riley, Noel Brady and all the others who have helped me to improve the preliminary version of these notes. Many thanks are due to the Centre de Recerca Matemàtica, and in particular to the organisers José Burillo and Enric Ventura, for their invitation to give this course, and for ensuring the smooth running of the event. And of course one is always grateful to the members of the audience for making the workshop a lot of fun.

In the first chapter, we shall see how the Cayley graph of a finitely generated group gives a geometric object providing a language in which to talk about many of the properties of the group. Geometric group theory studies properties of the Cayley graphs of groups.

In the second chapter we describe van Kampen's diagrams which provide a method for visualising relations in presentations of groups. We show how to obtain such diagrams and their dual pictures. We shall give some generalisations and applications to free products and HNN extensions.

Small cancellation theory gives a method of working with certain restricted forms of finite presentations; this is studied in Chapter 3. When a finite presentation satisfies certain easily verifiable conditions, the word problem is solvable in a particularly simple way. This theory has its origins in Dehn's original work, and led to Gromov's definition of word hyperbolic groups.

In Chapter 4 we give some details about quasi-isometries and show that a quasi-isometry of Cayley graphs preserves the property of being finitely presented, the property of having a solvable word problem, and the type of isoperimetric inequality satisfied. We shall describe some properties of word hyperbolic groups.

A certain method for obtaining lower bounds for isoperimetric inequalities is described in Chapter 5. This has an application to nilpotent groups, where this polynomial (of degree c=nilpotency class) bound can be combined with the $((c+1)$-degree) polynomial upper bound (see Chapter 4 of Tim Riley's notes) in the case of free nilpotent groups.

Finally we show how to obtain a polynomial isoperimetric inequality for certain normal subgroups of hyperbolic groups which are cyclic extensions. This applies to certain examples of Noel Brady.

Unfortunately time did not permit the covering of other topics, in particular Weinbaum's proof [30] of the conjugacy problem for alternating knots and links, Gromov's version of small cancellation theory (see Ollivier's paper [26]), and Klyachko's [19] work on the Kervaire conjecture (see also [12]).

Chapter 1

Dehn's Problems and Cayley Graphs

We shall suppose known the basic definition and properties of a free group — some of the many references available for this, in particular some favouring a geometric approach, are the books by Magnus, Karrass and Solitar [21], Lyndon and Schupp [LS], Bridson and Haefliger [6], Ghys and de la Harpe [14] and Hatcher [17].

We shall use $F(\mathcal{A})$ to denote the free group on $\mathcal{A}$. Let $R \subset F(\mathcal{A})$, and let $\langle\langle R \rangle\rangle$ denote the subgroup normally generated by R, i.e. the intersection of all normal subgroups which contain R. This is of course a normal subgroup, and it is not hard to see that it can be described as:

$$\langle\langle R \rangle\rangle = \{\prod_{i=1}^{M} p_i r_i{}^{\epsilon_i} p_i{}^{-1} \mid \forall M \in \mathbb{N}, \forall p_i \in F(\mathcal{A}), \forall r_i \in R, \forall \epsilon_i = \pm 1\}.$$

Let Γ be a group, and $\mathcal{A}$ a generating set for Γ. In the usual naïve sense, this means that $\mathcal{A}$ is a subset of Γ. (In this sense, the trivial group could only have one element in a generating set.) Here this will mean that there is a surjective homomorphism $\Phi : F(\mathcal{A}) \twoheadrightarrow \Gamma$. (Thus the trivial group can have a large generating set.) Any word w in (i.e. finite product of) the generators and their inverses thus represents an element of Γ; for the length of w we write $\ell(w)$, meaning the number of generators and their inverses appearing in the product. The obvious shortenings of this product, meaning the removal of subwords of the form aa^{-1} and $a^{-1}a$ for $a \in \mathcal{A}$ are called *reductions*, and the word is reduced if none are possible.

The kernel $\ker \Phi$ is a normal subgroup; if $R \subset F(\mathcal{A})$ is a subset which normally generates $\ker \Phi$, i.e. $\langle\langle R \rangle\rangle = \ker \Phi$, then we say that R is a set of relators for Γ with respect to the generating set $\mathcal{A}$. Such a set R always exists — it suffices to take $R = \ker \Phi$. What is more interesting is to try to obtain, if possible, a *finite* set R, or if not, some recursive or "systematic" set R. The corresponding presentation of Γ is written $\langle \mathcal{A} \mid R \rangle$ or $\langle \mathcal{A}; R \rangle$, meaning that the map $\mathcal{A} \to \Gamma$ induces an isomorphism $F(\mathcal{A})/\langle\langle R \rangle\rangle \to \Gamma$. The group Γ is finitely generated if it has a presentation with $\mathcal{A}$ finite, and is finitely presentable (or finitely presented) if it has a presentation with both $\mathcal{A}$ and R finite. We shall always assume that the words in R are cyclically reduced, as their cyclic reduction does not change the normal subgroup that is generated.

In 1912, Max Dehn [10] (this is available in an English translation, thanks to Stillwell) posed the three algorithmic problems for finitely presentable groups at the base of combinatorial group theory. It is worth noting that he did this well before Turing's and Gödel's work, though in the spirit of Hilbert's problems. It was not until the 1950s that it was proved (by Novikov and Boone) that in general such algorithms do not exist. Here are the three problems in their original formulation (Stillwell's translation [10, pages 133–134]) (see Figure 1.1 for the original):

The Word Problem: An element of the group is given as a product of generators. One is required to give a method whereby it may be decided in a finite number of steps whether this element is the identity or not.

The Conjugacy Problem: Any two elements S and T of the group are given. A method is sought for deciding the question whether S and T can be transformed into each other, i.e. whether there is an element U of this group satisfying the relation $S = UTU^{-1}$.

The Isomorphism Problem: Given two groups, one is to decide whether they are isomorphic or not (and further whether a given correspondence between the generators of one group and elements of the other group is an isomorphism or not).

geben. Man soll eine Methode angeben, um mit einer end-
lichen Anzahl von Schritten zu entscheiden, ob dies Element
der Identität gleich ist oder nicht.

   2. *Das Transformationsproblem:* Irgend zwei Elemente S
und T der Gruppe sind gegeben. Gesucht wird eine Methode
zur Entscheidung der Frage, ob S und T ineinander transfor-
miert werden können, d. h. ob es ein Element U der Gruppe
gibt, welches die Relation befriedigt:

$$S = UTU^{-1}.$$

   3. *Das Isomorphieproblem:* Zwei Gruppen sind gegeben,
man soll entscheiden, ob sie isomorph sind oder nicht (und,
des weiteren, ob eine gegebene Zuordnung der Erzeugenden
der einen Gruppe zu Elementen der andern Gruppe eine iso-
morphe Zuordnung ist oder nicht).

Figure 1.1: Dehn's three decision problems

An extremely efficient solution exists for certain finite presentations which have a *Dehn Algorithm*. We say that a presentation has such an algorithm when any word $w \in \langle\langle R \rangle\rangle$ always contains more than half of some relator (considered cyclically): i.e. w is a word in the generators of the form $Ur'V$ and there is some (cyclic conjugate of some) $r \in R \cup R^{-1}$ such that $r = r'r''$ and $\ell(r'') < \ell(r')$. If this is the case, then the group element represented by the subword r' is equal in the group to the element represented by r''^{-1}, and replacing r' by this shorter word reduces the length of w. Continuing in this way, w is trivial if and only if this procedure of looking for one of the finite number of long subwords of the cyclic conjugates of the relators, and replacing it by shorter word to give an element equal to w in the group, eventually leads to the empty word. If a finite presentation has this property, then the group presented is word hyperbolic, and any word hyperbolic group has such a finite presentation [2] (this was originally pointed out by Jim Cannon — in fact it has such a presentation with respect to any finite generating set: it suffices to add enough relators, see for instance [2]).

The first two problems describe a property of a finitely presentable group: if there is such an algorithm for the finite presentation $\mathcal{P}_1$ of Γ, and $\mathcal{P}_2$ is another finite presentation of Γ, then there is algorithm to solve the problem over $\mathcal{P}_2$ (as one can find an isomorphism between the presentations when one knows that one exists, by simply enumerating all Tietze transformations).

The third problem is so badly unsolvable in general that it is impossible to give an algorithm to recognize presentations of the trivial group. This is despite the fact that there is a procedure to enumerate all presentations of the trivial group (via Tietze transformations).

There is an obvious enumeration of $\langle\langle R \rangle\rangle$ by using the usual diagonal method on the lists of different numbers of conjugates of elements of $R^{\pm 1}$, ordered by the list of conjugating elements $p_i \in F(\mathcal{A})$. The hard part of the word problem resides in detecting words which represent *non-trivial* elements of Γ.

Given the enumeration procedure described above, if we know that the word w does indeed represent the trivial element in the presentation, the expression for w as a product of conjugates of relators $w = \prod_{i=1}^{M} p_i r^{\epsilon_i} p_i^{-1}$ can be found, where $\epsilon_i = \pm 1$ and $M \in \mathbb{N}$. The smallest such number M is called the *area* of w. The function $\delta_{\mathcal{P}} : \mathbb{N} \to \mathbb{N} : \delta_{\mathcal{P}}(n) = \max_{\{w \in \langle\langle R \rangle\rangle, \ell(w) \leq n\}} Area_{\mathcal{P}}(w)$ is called the *Dehn function* of the presentation $\mathcal{P}$. An *isoperimetric inequality* for the presentation is a function $f : \mathbb{N} \to \mathbb{R}$ such that for all $n \in \mathbb{N}$, $\delta_{\mathcal{P}}(n) \leq f(n)$. We shall study the dependence of these functions on the actual presentation later.

Theorem 1.1. *A finite presentation satisfies a recursive isoperimetric inequality if and only if it has a solvable word problem.*

Proof. If the word problem is solvable, then for each $n \in \mathbb{N}$, and each word w of length n, it is possible to decide whether or not w lies in $\langle\langle R \rangle\rangle$. If it does, then the enumeration procedure above eventually gives some expression for w as a product of conjugates of relators. In this way, examining all words of length at most n, this gives an upper bound for the Dehn function $\delta_{\mathcal{P}}(n)$ as required.

If a recursive function f bounding the Dehn function is known, and $w \in F(\mathcal{A})$ of length n is given, then calculate $f(n)$. It remains to calculate all products of at most $f(n)$ conjugates of the relators and their inverses. *A priori* the lengths of the conjugating elements which need to be tried are not bounded, and so all elements of $F(\mathcal{A})$ must be tried. We shall see later in Theorem 2.2, using van Kampen diagrams, that it suffices to check conjugating elements of length at most $f(n) \max_{r \in R} \ell(r) + \ell(w)$, so that there is a finite number of combinations which must be checked. This is the only step which is not immediate and it is best seen from the diagrams which are to be introduced in Chapter 2. □

In the 1980s Gromov introduced a class of groups generalising discrete groups of isometries acting cocompactly on hyperbolic spaces. It is an interesting exercise to read Gromov's papers about hyperbolic groups alongside Dehn's articles about the conjugacy problem for surface groups (for instance in Stillwell's translation, see [10]). There are several equivalent definitions of the class of hyperbolic groups, some of which we shall explore later. One definition is in terms of area functions:

Definition 1.2. A finitely presentable group Γ is *word hyperbolic* if it has a finite presentation which satisfies a linear isoperimetric inequality.

In fact a group is word hyperbolic if and only if it satisfies a sub-quadratic isoperimetric inequality [4, 27]. Also, as we noted earlier, a group is word hyperbolic if and only if it has a finite presentation which has a Dehn algorithm (see for instance [2]). It is easy to solve the word and conjugacy problems for finitely generated abelian groups, or free groups with respect to free generating sets. The first case gives a quadratic isoperimetric inequality, the second a linear one.

The idea of representing a group by a graph goes back to Cayley, though he only uses them in the context of finite groups. Dehn extends the ideas to infinite groups and uses them (*Gruppenbilder*) intensively to study fundamental groups of closed compact surfaces.

Definition 1.3 (Cayley graph). Let $\mathcal{A}$ be a generating set for the group Γ. The Cayley graph of Γ with respect to $\mathcal{A}$, written $Cay^1(\Gamma, \mathcal{A})$, has a vertex for each element $g \in \Gamma$, and for each such vertex g_1, and each $a \in \mathcal{A}$, there is an oriented labelled edge from the vertex g_1 to the vertex g_2 if and only if $g_2 = g_1.a$ in Γ (which we write $g_2 =_\Gamma g_1 a$). Notice that it may be that $g_1 =_\Gamma g_2$ when the generator a represents the identity element of Γ. The fact that $\mathcal{A}$ is a generating set means that this graph is connected. Assign length 1 to each edge to consider $Cay^1(\Gamma, \mathcal{A})$ as a metric space, where the distance $d_{\mathcal{A}}(v, v')$ between the points v, v' is the length of the shortest path between them.

On Γ this defines an integer-valued metric on Γ called the word-metric (it depends on the choice of generating set). As is usual in the context of labelled oriented graphs, to a path between vertices in the Cayley graph is associated a word in the free group $F(\mathcal{A})$ (this word may be unreduced). The word is obtained by writing the letter corresponding to the label on each edge in the order traversed, with the exponent ± 1 according to whether the direction of the path agrees with $(+1)$ or is opposite to (-1) the orientation of the edge.

I represent this by a diagram, the lines of which were red and black, and they

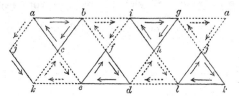

will be thus spoken of, but the black lines are in the woodcut continuous lines, and the red lines broken lines: each face indicates a cyclical substitution, as shown by the arrows. The figure should be in the first instance drawn with the arrows, but without the letters, and these may then be affixed to the several points in a perfectly arbitrary manner; but I have in fact affixed them in such wise that the group given

Figure 1.2: From Cayley's paper [7]: a Cayley graph of the alternating group A_4, with presentation $\langle x, y \mid x^3, y^3, (xy)^3 \rangle$ (the vertices a, j, k are repeated).

Some elementary properties of Cayley graphs:

- The group Γ acts freely on the left on $Cay^1(\Gamma, \mathcal{A})$ by isometries with respect to the distance $d_{\mathcal{A}}$, and transitively on the set of vertices.

- The Cayley graph $Cay^1(\Gamma, \mathcal{A})$ of a finitely generated group is a covering space of the 1-complex $K(\mathcal{A})$, with one vertex and $|\mathcal{A}|$ edges (each forming a loop), whose fundamental group is $F(\mathcal{A})$. We can regard $K(\mathcal{A})$ as $Cay^1(1, \mathcal{A})$, a Cayley graph of 1, the trivial group.

- The usual correspondence between covering spaces and subgroups of the fundamental groups, says that when $\langle \mathcal{A} \mid R \rangle$ is a presentation (finite or infinite) of the group Γ, the Cayley graph $Cay^1(\Gamma, \mathcal{A})$ is the cover of $K(\mathcal{A})$ corresponding to the normal subgroup $\langle\langle R \rangle\rangle$ of $F(\mathcal{A})$, the fundamental group of $K(\mathcal{A})$.

- In fact, if Γ' is a normal subgroup of Γ, then the quotient space $Cay^1(\Gamma, \mathcal{A})/\Gamma'$ is a Cayley graph of Γ/Γ'.

The diagram has the property that every route, leading from any one letter to itself, leads also from every other letter to itself; or say a route leading from a to a, leads also from b to b, from c to c, ..., from l to l; and we can thus in the diagram speak absolutely (that is, without restriction as to the initial point) of a route as leading from a point to itself, or say as being equal to unity; *it is in virtue of this property* that the diagram gives a group.

Figure 1.3: Cayley's description of the group acting on its graph.

- Fixing some vertex v of $Cay^1(\Gamma, \mathcal{A})$ as a base point (for instance the vertex corresponding to the identity element 1), the word $w \in F(\mathcal{A})$ defines a unique path γ_w based at v. The word represents the identity element of Γ if and only if the path γ_w is a loop (note Cayley's remark on this, Fig. 1.3).

- To solve the word problem, it suffices to build the Cayley graph, or at least, to give an algorithm which, given a word of length n, constructs the ball of radius $n/2$ about the identity element in the Cayley graph, which is enough of it to see whether or not the word w labels a loop or not (this is basically the Todd–Coxeter algorithm).

Here are three essential, elementary, examples of Cayley graphs. For more examples of Cayley graphs of two generator groups, seen as covering spaces, see [17, p. 58].

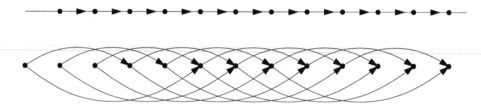

Figure 1.4: Two Cayley graphs for $\mathbb{Z}$ with generating sets $\{1\}$ and $\{3,5\}$.

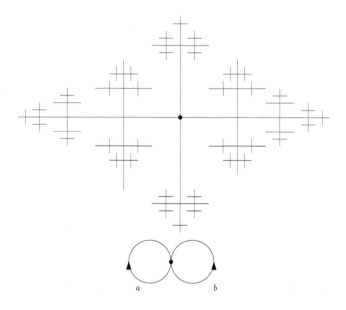

Figure 1.5: The Cayley graph of a free group on two generators with respect to a free basis; the a edges are horizontal, oriented from left to right, the b edges vertical oriented upwards.

Chapter 2

Van Kampen Diagrams and Pictures

We now introduce the principal tool we shall use here for examining the word problem. These are diagrams introduced by Egbert van Kampen in 1933 [29].

Definition 2.1 (van Kampen or Dehn diagram). Let $\mathcal{P} = \langle \mathcal{A} \mid R \rangle$ be a (usually finite) presentation for the group Γ. As is usual, we shall suppose that the relations in R are cyclically reduced. Let R^C denote the cyclic closure of R, which is the set of all cyclic conjugates of elements of R and their inverses:

$$R^C = \{(p^{-1}rp)^{\pm 1} \mid p \text{ an initial segment of } r \in R\}.$$

Let $\mathcal{D}$ be a finite, connected, oriented, based, labelled, planar graph where each oriented edge is labelled by an element of $\mathcal{A}$. The base point lies on the boundary of the unbounded region of $\mathbb{R}^2 - D$. Suppose in addition that for each bounded region (face) F of $\mathbb{R}^2 - D$, the boundary ∂F (of the closure of F) is labelled by a word in R^C. This word is obtained by reading the labels on the edges as they are traversed, starting from some vertex on the boundary of F, in one of the two possible directions. Each label on the edge traversed is given a ± 1 exponent according to whether the direction of traversal coincides with, or is opposite to, the orientation of the edge. The choice of direction and starting point alters the word read by inversion and/or cyclic conjugation. The *boundary word* of the diagram D is the word w read on the boundary of the unbounded region of $\mathbb{R}^2 - D$, starting from the base vertex. Then we say that $\mathcal{D}$ is a *van Kampen diagram* for the boundary word w over the presentation $\mathcal{P}$.

The diagram can also be viewed as a 2-complex, with a 2-cell attached to the graph (viewed as a 1-complex) for each bounded region. This constructs a combinatorial 2-complex, as we shall see below.

Usually we can suppose that w is a freely reduced word, though probably not cyclically reduced.

In 1933, Egbert van Kampen [29] defined his diagrammatic method of considering which words represent the identity element in the group given by a finite presentation. There he (essentially) stated the following result.

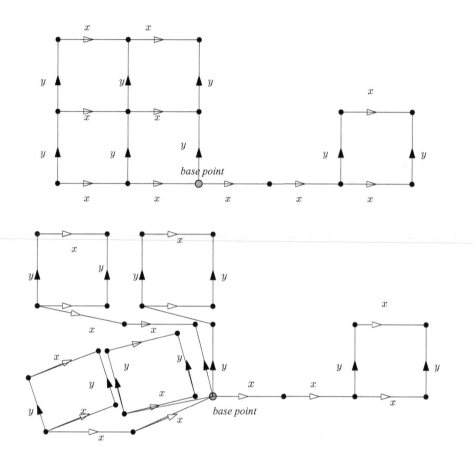

Figure 2.1: A diagram for $w = x^3yx^{-1}y^{-1}x^{-2}y^2x^{-2}y^{-2}x^2$ over the presentation $\langle x,y \mid xyx^{-1}y^{-1}\rangle$, and the same diagram for w deconstructed as $w = x^2rx^{-2}.yx^{-1}rxy^{-1}.yx^{-2}rx^2y^{-1}.x^{-1}rx.x^{-2}rx^2$, where $r = xyx^{-1}y^{-1}$

Theorem 2.2. *Let* $\mathcal{P} = \langle \mathcal{A} \mid R\rangle$ *be a presentation of the group* $\Gamma = \Gamma(\mathcal{P})$.

1) *If* $w \in \langle\langle R\rangle\rangle$, *i.e.* $w =_{\mathcal{P}} 1$, *then there is a van Kampen diagram for* w *over* $\mathcal{P}$.

2) *If* $\mathcal{D}$ *is a van Kampen diagram for* w *over* $\mathcal{P}$, *then* $w =_{\mathcal{P}}$.

We shall establish something stronger, where we may allow bounded regions of the planar graph to have labels which are not freely (nor cyclically) reduced. This more general form will be useful when dealing with cancelling faces in Definition 2.7.

Proposition 2.3. *Let* $\mathcal{D}$ *be a finite, connected, oriented, based, labelled, planar graph where each oriented edge is labelled by an element of* $\mathcal{A}$. *Suppose in addition that* *the bounded regions of $\mathbb{R}^2 - D$ are labelled by words whose freely reduced forms lie in $\langle\langle R\rangle\rangle$, and that the boundary label is w. Then a finite, connected, oriented, based, labelled, planar graph D' can be obtained from D such that all bounded regions are labelled by cyclically reduced words in $\langle\langle R\rangle\rangle$, and the boundary label is the reduced word corresponding to w.*

Proof. 1) We shall assume that w is given as a freely reduced word. Write $w = \prod_{i=1}^{M} p_i r_i^{\epsilon_i} p_i^{-1}$, with $r_i \in R$, $\epsilon_i = \pm 1$, $p_i \in F(\mathcal{A})$. This is an equality in the free group $F(\mathcal{A})$.

If $M = 1$, then draw in the plane a circle subdivided into $\ell(r_1)$ segments (or a regular polygon with $\ell(r_1)$ sides), and add an arc outside the circle at one of the vertices; subdivide the arc into $\ell(p_i)$ segments. Orient and label each of the edges appropriately. This is a van Kampen diagram for w (or rather for a word freely equal to w). We show below how to alter the diagram to obtain a diagram such that the boundary word is freely reduced.

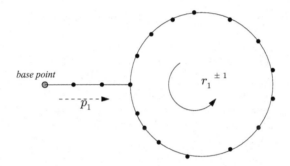

Figure 2.2: The case $M = 1 : w$ is the conjugate of a single relation

The general case: Draw M copies of the circle plus arc as before, all based at the same base point in the plane. Subdivide the arcs and the circles, and orient and label them appropriately.

When we assume that the relators in R are assumed to be cyclically reduced words, we can also ensure that the labels on all the boundaries of the regions are freely reduced words — a 1-dimensional reduction involving cancelling 1-cells.

If there is a vertex in the diagram other than the base vertex, which has valency one, then removing this vertex and the incident edge changes the label on the region having this edge on its boundary by a free reduction. If the valency-one vertex is the base vertex, then removing this edge and vertex would correspond to a *cyclic* reduction of the boundary word w, which in general we do not allow, as we regard this word being fixed (up to free reduction).

Suppose that there are two edges $e_1 = (v, v_1)$, $e_2 = (v, v_2)$ emanating from the same vertex $v \in \mathcal{D}$, both edges labelled by the same letter $x \in \mathcal{A}$, with the same orientation with respect to v, and such that e_1 and e_2 are adjacent edges on the boundary of some face f of $\mathcal{D}$ (see Figure 2.4). Identify the two edges e_1

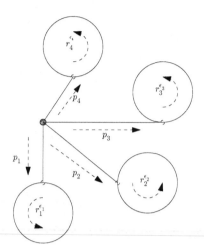

Figure 2.3: First step in the construction of a van Kampen diagram: general case.

and e_2, identifying the vertices v_1 and v_2. This changes the face f to the face f', with two fewer edges, and in the label on the boundary of this face, the cancelling letters $x^{-1}x$ are removed; all other face labels are unchanged.

Van Kampen says: *the two 1-cells can be brought into coincidence by a deformation without any other change in the complex*. There is however a problem, as there are four cases to consider:

Case 1: $v_1 \neq v \neq v_2$ and $v_1 \neq v_2$. Case 2: $v_1 = v \neq v_2$.
Case 3: $v_1 = v = v_2$. Case 4: $v_1 = v_2 \neq v$.

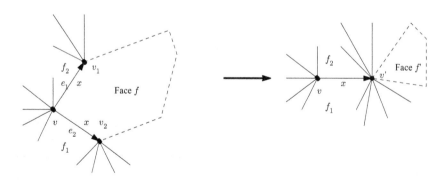

Figure 2.4: Case 1: Edge identification when $v_1 \neq v \neq v_2$; here the face f is bounded. The identification can be realised by a map of the plane to the plane which essentially collapses a triangle onto an edge. Only one boundary label is affected, and that by a free reduction.

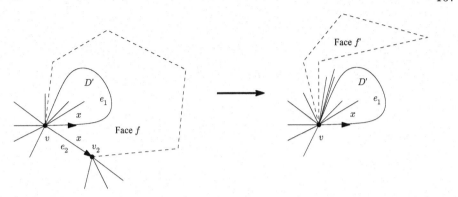

Figure 2.5: Case 2: $v_1 = v \neq v_2$. Again the identification can be realised by a map from the plane to the plane which collapses a triangle onto an edge, and only one boundary label is affected, as before.

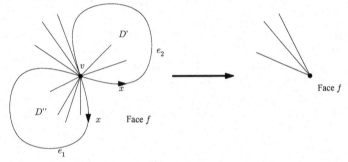

Figure 2.6: Case 3: $v_1 = v = v_2$. Here the identification of e_1 with e_2 cannot be realised while at the same time resting in the plane. However, we can realise the identification as a graph in the plane, together with a graph in a 2-sphere, made up of the two bounded regions D', D'' of the plane bounded by the edges e_1 and e_2. This 2-sphere is attached to the rest of the diagram at the vertex v. The 2-sphere is then discarded, leaving a diagram with fewer faces, and with the same labels on the remaining faces, except the face f where a free reduction, cancelling the adjacent letters x and x^{-1}, has been performed.

2) Suppose now that $\mathcal{D}$ is a van Kampen diagram, based at the vertex v, with M bounded faces. Let f be a bounded face which meets the unbounded region of the plane in a non-empty set B, containing an edge e, and let v_f be the initial vertex of e. Let r be the boundary label on f when read from v_f. Let $\gamma \subset \partial D$ be a simple arc in the boundary ∂D from v to v_f, with label p.

Removing the interior of the edge e from $\mathcal{D}$ gives a van Kampen diagram $\mathcal{D}'$ with $M - 1$ bounded regions, possibly with some vertices of valency one. In the unbounded region of the diagram $\mathcal{D}'$, join to the base point v an arc labelled p leading to a disk with subdivided boundary labelled $r^{\pm 1}$. Continuing in this way,

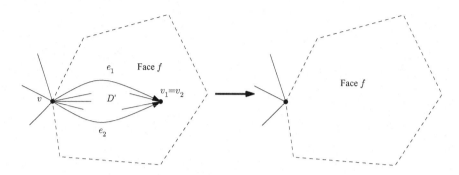

Figure 2.7: Case 4: $v_1 = v_2 \neq v$. Here the entire subdiagram D' enclosed by the two edges e_1 and e_2 is removed. Identifying the edges e_1 and e_2 produces a 2-sphere attached to the plane along the edge $e_1 = e_2$.

the whole diagram can be "deconstructed" to give a bouquet of circles describing w as a product $w = \prod_{i=1}^{M} p_i r_i p_i^{-1}$ (as in Figure 2.3). Again the equality here is in the free group $F(\mathcal{A})$. Notice that the length of the words p_i is bounded by the number of edges in $\mathcal{D}$, which is bounded by $\ell(w) + M\rho$ where $\rho = \max_{r \in R} \ell(r)$. Alternatively, it is possible to find other words p_i of length bounded by $\ell(w)/2$ plus the length of the shortest path in D from the i-th face to the boundary. $\square$

It is important to underline the following aspects of the form of a van Kampen diagram. Regarding the diagram as a planar 2-complex, it is a collection of disjoint closed topological 2-cells joined by arcs (and vertices): removing the closures of these 2-cells from the diagram leaves a collection of trees. There is a retraction of the diagram onto a tree, realised by retracting each of the disc components to a point.

In fact the first part of the above proof establishes something stronger, where we may allow bounded regions of the planar graph to have labels which are not freely (nor cyclically) reduced. This will be useful when describing reduction of diagrams (see Definition 2.7).

Proposition 2.4. *Let $\mathcal{D}$ be a finite, connected, oriented, based, labelled, planar graph where each oriented edge is labelled by an element of $\mathcal{A}$. Suppose in addition that the bounded regions of $\mathbb{R}^2 - D$ are labelled by words whose freely reduced forms lie in $\langle\langle R \rangle\rangle$, and that the boundary label is w. Then a finite, connected, oriented, based, labelled, planar graph D' can be obtained from D such that all bounded regions are labelled by cyclically reduced words in $\langle\langle R \rangle\rangle$, and the boundary label is the reduced word corresponding to w.*

Definition 2.5 (The presentation complex and the Cayley complex). Let $\mathcal{P} = \langle \mathcal{A} \mid R \rangle$ be a group presentation. The standard 2-complex $\mathcal{K}^2(\mathcal{P})$ associated to the presentation $\mathcal{P}$, consists of:

a single vertex v, and an oriented 1-cell $e_i^{(1)}$ for each $a_i \in \mathcal{A}$, labelled a_i;

a 2-cell $e_j^{(2)}$ for each $r_j \in R$.

The 1-cells are attached to the 0-skeleton in the only way possible. Thus the fundamental group of the 1-skeleton is the free group on the set $\mathcal{A}$. The 2-cell $e_j^{(2)}$ is attached to the 1-skeleton via a map identifying its boundary with a loop in the 1-skeleton corresponding to the word r_j. The resulting space is given the quotient topology. Giving all 1-cells unit length, and viewing the 2-cell $e_j^{(2)}$ as a regular euclidean polygon with $\ell(r_j)$ sides each of unit length, gives an additional piecewise euclidean structure to the complex. (When there are relations of length 1 or 2, use a disk with circumference length 1 or 2; such relators can of course be easily avoided.)

The Seifert–van Kampen theorem (in the simpler case when one component is simply connected) tells us that $\pi_1(\mathcal{K}^2(\mathcal{P}), v) = \Gamma(\mathcal{P})$.

The Cayley complex of the group $\Gamma(\mathcal{P})$ with respect to the presentation $\mathcal{P}$ is the universal covering space of $\mathcal{K}^2(\mathcal{P})$. This is obtained from $Cay^1(\Gamma, \mathcal{A})$ by adding 2-cells: for each $g \in \Gamma$, a 2-cell for each r_j is added, based at the vertex corresponding to g (i.e. a lift based at each vertex g of each attaching map). Notice that for a relation r, which is a proper power, say $r = s^q$, the vertex $g \in Cay^2(\Gamma, \mathcal{A})$ is in the boundary of q attached 2-cells labelled r, as there are q lifts of the attaching map which pass through this vertex, one based there, the others based at $gs^{q'}$ for $1 \leq q' < q$.

Examples.

1) The presentation $\langle x, y \mid xyx^{-1}y^{-1} \rangle$ gives a standard complex homeomorphic to the torus $S^1 \times S^1$, whose Cayley complex (as a P.E. complex) is (isometric to) the plane $\mathbb{R} \times \mathbb{R}$. The Cayley graph is the 1-skeleton, viewed as the set of points with at least one integer coordinate, and the vertices are those points with two integer coordinates.

2) The presentation $\mathcal{P} = \langle a, b \mid a^2, b^2 \rangle$ gives a standard complex homeomorphic to two projective planes joined at a point. The Cayley complex $Cay^2(\mathcal{P})$ is homeomorphic to an infinite collection of spheres indexed by $\mathbb{Z}$, where the north pole of the i-th sphere is joined to the south pole of the $(i+1)$-st sphere.

Using this complex, we can give a topological derivation of van Kampen diagrams. This leads to diagrams which are essentially dual to van Kampen diagrams, and in their full generality are known as transversality diagrams. In this context they were introduced by Colin Rourke (see also Fenn's book [11]). As is customary, we take for granted many results about transversality. The advantage of this alternative treatment is that the diagrammatic method is based on topology rather than combinatorics, though this is perhaps mostly a matter of taste.

Definition 2.6 (Pictures). Consider the presentation $\mathcal{P} = \langle \mathcal{A} \mid R \rangle$, the standard complex $\mathcal{K}^2(\mathcal{P})$ and $w \in \langle\langle R \rangle\rangle$. Then w defines a path γ_w in the 1-skeleton $\mathcal{K}^2(\mathcal{P})^{(1)}$ which is null-homotopic in $\mathcal{K}^2(\mathcal{P})$. This means that there is a map $f : (D, \partial D) \to (\mathcal{K}^2(\mathcal{P}), \mathcal{K}^2(\mathcal{P})^{(1)})$ such that $f|_{\partial D} = \gamma_w$, and the homotopy lifting property says that f lifts to a map $(D, \partial D) \to (Cay^2(\mathcal{P}), Cay^1(\mathcal{P}))$.

  • After a homotopy (relative to ∂D), we can suppose that, for each relator r_j, f is transverse to the centre $\hat{e}_j$ of the corresponding 2-cell e_j. This means that $f^{-1}(\hat{e}_j)$ is a finite set of points in int D, and that there is a disjoint set of open neighbourhoods $V(\alpha_{j,k})$ of the points $\alpha_{j,k} \in f^{-1}(\hat{e}_j)$ such that the restriction of f to each neighbourhood is a homeomorphism into a neighbourhood of $\hat{e}_j$ in int(e_j).

  • As $f(D - \bigcup_{j,k} V(\alpha_{j,k})) \subset \mathcal{K}^2(\mathcal{P}) - \bigcup_j \hat{e}_j$, and this latter space retracts onto the 1-skeleton $\mathcal{K}^2(\mathcal{P})^{(1)}$, after a further homotopy of f (fixing the boundary ∂D), we can suppose that $f(D - \bigcup_{j,k} V(\alpha_{j,k})) \subset \mathcal{K}^2(\mathcal{P})^{(1)}$.

  • Now, by a further homotopy (fixing the boundary), we can make $f : D - \bigcup_{j,k} V(\alpha_{j,k}) \to \mathcal{K}^2(\mathcal{P})^{(1)}$ transverse to the mid-points $\hat{e}_i$ of the 1-cells $e_i{}^{(1)}$. This means that $f^{-1}(\hat{e}_i)$ is a finite set of properly embedded arcs and loops in $D - \bigcup_{j,k} V(\alpha_{j,k})$. Moreover each arc/loop has a transverse orientation and label coming from the orientation and label in a neighbourhood of $\hat{e}_i$ in $e_i{}^{(1)}$. (Here we should note that during the construction of the complex $K(\mathcal{P})$, the attaching maps of the 2-cells should have been made transverse to the points $\hat{e}_i$.)

The *picture* corresponding to f is the disc D together with the collection of subdisks (or "fat vertices") $V_{j,k}$ and the embedded loops and arcs. Each arc and loop is transversely oriented and labelled by some $x \in \mathcal{A}$, inducing labels r_j on the boundary of each $V_{j,k}$, and a label w on ∂D, when read from appropriate points, and in an appropriate direction.

The picture can be thought of as constructed from copies of small discs ("fat vertices") with protruding "legs", each of which has a transverse orientation and a label from $\mathcal{A}$: each disc corresponds to some $r \in R$, and the labels on the edges, read from some base point, with exponents ± 1 according to the orientation spell out the word r (Roger Fenn calls these "spiders and anti-spiders" in [11]).

Notice that there may be free loops in the picture: i.e. there may be a simple loop in D which maps to a point $\hat{e}_i{}^{(1)}$. Such a loop, and all of the picture in the subdisk of D bounded by it, can be removed to obtain a simpler picture. In fact the interior of the subdisk corresponds to a picture on a 2-sphere.

It is easy to see that pictures are basically dual to diagrams: to obtain a picture from a diagram: surround the diagram in the plane by a big circle. Insert a small circle in the interior of each compact region. Dual to an edge separating two faces of the diagram, insert an edge joining the added circles in each face. Label and transversely orient these added edges according to the orientation of the original edge.

Given a picture, remove all free loops. Around each small disc with n legs, draw a polygon with n sides. The sides are labelled and oriented according to the label and (transverse) orientation of the legs. For each arc in the picture, the sides

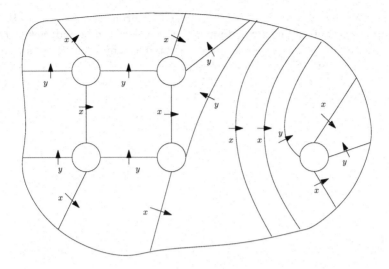

Figure 2.8: A picture for $w = x^3yx^{-1}y^{-1}x^{-2}y^2x^{-2}y^{-2}x^2$, cf. Figure 2.1.

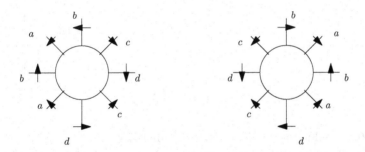

Figure 2.9: A spider and an anti-spider for the relation $r = aba^{-1}b^{-1}cdc^{-1}d^{-1}$.

of polygons occurring at the two ends of the arc are identified: the danger here is that this may lead to a non-planar diagram (thus the word "basically" above).

There is an obvious simplification that can be performed on diagrams and on pictures:

Definition 2.7. Let D be a van Kampen diagram. Let F_1 and F_2 be distinct compact regions of D, such that there is at least one edge e in the intersection $\partial F_1 \cap \partial F_2$.

For $i = 1, 2$, let r_i be the label read on the boundary ∂F_i starting from the initial vertex of e, and reading in the direction induced by the orientation of e (see the example in Figure 2.10). If $r_1 = r_2$ (i.e. identical as words in $F(\mathcal{A})$), then the diagram is said to be *unreduced*, and if no such pair of faces exists, then the diagram is said to be *reduced*.

Removing the edge e from the unreduced diagram gives the possibility of performing a series of foldings which identify the rest of the boundaries of F_1 and F_2, while leaving unaffected the remainder of the diagram (as in Proposition 2.4). In this way a diagram with two fewer regions is obtained (maybe there are many fewer regions after the folding has finished). As an exercise in this dual method, we now describe the reduction procedure in the world of pictures.

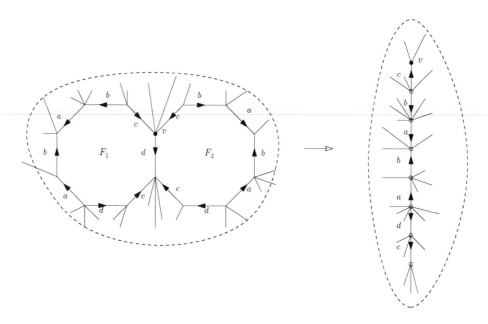

Figure 2.10: A reduction in a van Kampen diagram; here the common edge e, labelled d, is removed, then the other edges are in turn identified.

Translating into the world of pictures, let P be a picture in which two small discs are joined by at least one arc, and such that the labels on the two small discs, starting from the label on one particular arc joining them, and in the direction induced by the transverse orientation on that arc, are the same. Then the picture can be altered to remove these two small disks as indicated in Figure 2.11.

We saw in the world of van Kampen diagrams that care has to be taken when there is more than one segment in the intersection of the boundaries of F_1 and F_2, and in this case a part of the diagram may have to be discarded in order to retain planarity. This problem translates in the pictures world to two collections of arcs joining the two small disks with possibly some subpicture trapped between them as in Figure 2.12. The reduction process can then lead to a subpicture unconnected with the rest of the picture. This subpicture can be discarded without affecting the boundary label, or the essential properties of the picture.

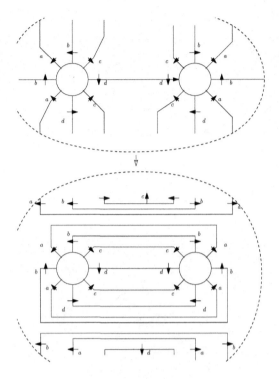

Figure 2.11: A reduction of cancelling spider and anti-spider.

There is much which can be said about reduced diagrams and the relationship with π_2 and the various definitions of "aspherical". In this topological version, it is clear that a map of a sphere into the Cayley complex leads to a picture on a sphere, using the same transversality and homotopy construction as was used for the map of a disc into the complex. (See [8] for a detailed discussion of some of the issues relating to this method of studying π_2, including the problems with the various definitions of *aspherical* used in [20]).

We have noted that when $w =_{\mathcal{P}} 1$, there is a diagram for w over $\mathcal{P}$, and there is clearly a reduced diagram, and in fact a smallest diagram, with the smallest number of compact faces, i.e. a diagram D such that $area(D) = area(w)$ is reduced. The area of a picture is naturally the number of small disks (fat vertices). There are certain presentations where for each word w representing the trivial element, there is a unique reduced diagram.

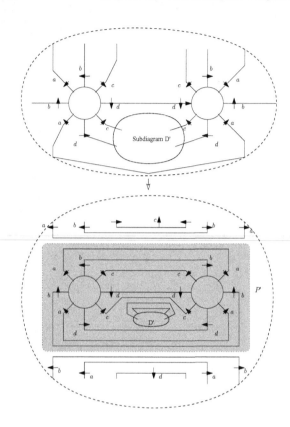

Figure 2.12: Reduction produces a disconnected picture: the subpicture P' can then be discarded.

Britton's Lemma and Collins' Lemma for HNN extensions via pictures

As usual, let $\langle \mathcal{A} \mid R \rangle$ be a presentation of a group Γ. Consider the HNN extension:

$$H = \Gamma *_C = \langle \Gamma, t \mid tct^{-1} = \phi(c), c \in C \rangle = \langle \mathcal{A}, t \mid R, tct^{-1} = \phi(c), c \in C \rangle$$

for some subgroup $C < \Gamma$ and some injective homomorphism $\phi : C \to \Gamma$. To ensure that this group is finitely presented, we suppose that $\mathcal{A}, R$ are finite, and that C is finitely generated.

Lemma 2.8 (Britton's Lemma). *Let $w = b_1 t^{\alpha_1} b_2 t^{\alpha_2} \ldots b_k t^{\alpha_k}$ be a non-empty re-duced word in $F(\mathcal{A}, t)$, where $b_i \in F(\mathcal{A})$, such that $\alpha_i \neq 0$ for $i < k$.*
*   If $w =_H 1$ and $\alpha_1 \neq 1$, then for some $i = 2, \ldots, k$, either $\alpha_{i-1} > 0 > \alpha_i$ and $b_i \in C$, or $\alpha_{i-1} < 0 < \alpha_i$ and $b_i \in \phi(C)$.*

(There are two such subwords b_i if sub-indices are considered modulo k.)

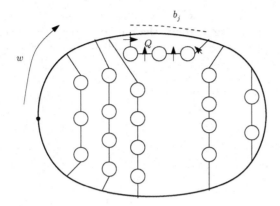

Figure 2.13: The proof of Britton's Lemma: suppress all edges except those labelled by t, and all relation disks except those labelled by $tat^{-1}\phi(a)^{-1}$. If there are rings formed by the t-edges then these can be removed as in the proof of Proposition 2.9 below. There is some "outermost region" bounded by t edges (here labelled Q) which does not contain the base point v, and this region meets the boundary in a segment labelled by the subword b_i of w. The labels on the two t-edges at either end of the b_j segment are opposite due to the consistency of the t orientation along the chain of t edges and relation discs. Thus $t^\epsilon b_i t^{-\epsilon}$ is a subword of w.

Once this result has been established, we can assume that elements in H are represented by words which do not contain "pinches". A *pinch* in a word on the generators for the HNN extension H is a subword of the form tbt^{-1} where $b \in C$, or of the form $t^{-1}bt$, where $b \in \phi(C)$ (a cyclic pinch is a pinch in the word viewed cyclically). Detecting the existence of such pinches in w depends upon having an algorithm to decide whether or not a word in the generators for H represents an element of $C \cup \phi(C)$. The problem of deciding, for a finite set of elements X in the group G, whether or not a word w lies in the subgroup generated by X, is known as the *generalised word problem*. A consequence of Britton's lemma is thus:

Proposition 2.9 (The word problem for HNN extensions). *The word problem is solvable for H if the word problem is solvable for Γ and the generalised word problem for C and for $\phi(C)$ in Γ is solvable.*

Proof. Notice that the solution of word problem for Γ is included in the solution of the generalised word problem for C. Let $w = b_1 t^{\alpha_1} b_2 t^{\alpha_2} \ldots b_k t^{\alpha_k}$ be a cyclically reduced word in $F(\mathcal{A}, t)$.

Case 1: There are no occurrences of t in w.

If $w =_H 1$, then there is a corresponding picture in which no t-edges meet the boundary. If there are no t-edges in the picture, then $w = 1$ in Γ, and the algorithm for the word problem in Γ tells us whether or not $w = 1$ in Γ. If there are t-edges in the picture, then they are joined to relation discs in such a way as to form rings.

An innermost such ring has a label w' in $F(\mathcal{A})$, and the sub-picture enclosed is a picture for $w' = 1$ in Γ. The word on the outside of this ring is either $\phi(w')$ or $\phi^{-1}(w)$, according to the orientation of the t-edges. In either case, $w' = 1$ in Γ if and only if $w' \in C$ and $\phi(w') = 1$ in Γ, or $w' \in \phi(C)$ and $\phi^{-1}(w') = 1$ in Γ. In both cases, an innermost such ring can be removed and replaced by a picture over Γ for the outside word ($\phi(w')$ or $\phi^{-1}(w')$) with no t-edges. In this way, after repetitions, if $w = 1$ in H and there are no t occurrences in w, then there is a picture for w over Γ (and thus the natural homomorphism of Γ into H induced by the map on the generators is in fact injective).

Case 2: There are occurrences of t in w: proceed by induction on the number of such t-occurrences.

Suppose there is an algorithm to decide triviality when a word has at most N t-occurrences. Suppose that w is a word with $N + 1$ t-occurrences. If $w = 1$ in H, then Britton's Lemma says that there is a subword of w of the form tb_jt^{-1} or $t^{-1}b_jt$ for some j, such that $b_j \in C$ or $b_j \in \phi(C)$ respectively. The algorithm to decide membership of the subgroups C and $\phi(C)$ in Γ decides whether or not such subwords exist in w, and when such a subword is found, it is replaced by $\phi(b_j)$ or $\phi^{-1}(b_j)$ respectively, giving a new word with fewer t-occurrences which is equal to w in H. Notice that here $\phi(b_j)$ is calculated via a semigroup homomorphism when C is given as a finitely generated subgroup of Γ, and $\phi(c_i)$ is given for each generator of C.   □

The conjugacy problem for HNN extensions can be studied in a similar way. This diagrammatic proof was given by Miller and Schupp [22]. Here the relevant lemma concerns annular pictures:

Lemma 2.10 (Collins' Lemma). *Let* $u = b_1t^{\beta_1}b_2t^{\beta_2}\ldots b_pt^{\beta_p}$, $v = d_1t^{\delta_1}d_2t^{\delta_2}\ldots d_qt^{\delta_q}$ *be cyclically reduced words in* $F(\mathcal{A}, t)$ *such that* $b_i, d_j \in F(\mathcal{A}) - \{1\}$, *and* $\beta_i, \delta_j \in \mathbb{Z} - \{0\}$, *except when* $u = b_1$ *or* $v = d_1$. *Suppose in addition that there are no pinches or cyclic pinches in* u *or* v.
If u *and* v *are conjugate in* $H = \langle \Gamma, t \mid tCt^{-1} = \phi(C) \rangle$, *then one of the following holds:*

- u, v *are words in* $F(\mathcal{A})$ *which are conjugate in* Γ.

- *There is a finite chain of words in* $F(\mathcal{A})$, $u = v_0, u_1, v_1, u_2, v_2, \ldots, u_k$, v_k, u_{k+1}, *such that* $v_i = \phi^{\pm 1}(u_i)$, *as group elements,* $u_i, v_i \in C \cup \phi(C)$, *and for each* $i = 0, \ldots k$, v_i *is conjugate to* u_{i+1} *by an element of* Γ.

- *both* u *and* v *contain* t *occurrences, and* $p = q$, $u = b_1t^{\beta_1}b_2t^{\beta_2}\ldots b_pt^{\beta_p}$ *is conjugate to some cyclic conjugate of* v *of the form*

$$d_it^{\delta_i}d_{i+1}^{\delta_{i+1}}\ldots d_qt^{\delta_q}d_1t^{\delta_1}\ldots d_{i-1}t^{\delta_{i-1}},$$

and the conjugation can be realised by an element of $C \cup \phi(C)$.

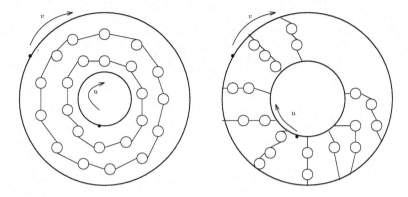

Figure 2.14: The two possible non-trivial conjugacy pictures for Collins' Lemma are annular pictures where the boundary components are labelled u and v (the first case of the statement corresponds to a picture without t-edges). Again suppress all edges except for t-edges, and suppose that there are some t-occurrences in u and/or v. Remove closed loops of t-edges which bound discs in the annulus as in the proof of 2.9 above. As there are no cyclic pinches, all chains of t-edges and relations go from one boundary component to the other, or form closed loops which are "parallel" to the boundary (i.e. go once round the annulus). These two types cannot coexist, and this leads to the second two cases of the statement.

The conjugacy problem is in general *not* solvable for HNN extensions, even when the conjugacy and word problems are solvable in Γ. The difficulty lies in the estimation of the number of conjugations in the second case of Collins' lemma.

If the generalized word problem is solvable for C and $\phi(C)$ in Γ, then cyclic pinches can be detected in words in $F(\mathcal{A}, t)$.

If after removal of these pinches from two words one of the words contains t-occurrences, then the other word must also have t-occurrences, and case 3 of Collins' Lemma applies.

Chapter 3

Small Cancellation Conditions

Certain conditions on the form of the relators in a presentation, and how they interact locally, can give strong restrictions on the properties of the group so presented. The so-called "small cancellation conditions" developed in the 1960s by Lyndon, Greendlinger and others quickly give isoperimetric inequalities. A full presentation is given in Chapter V of Lyndon and Schupp's book [20]. These conditions are also sufficiently generic to give results about "random" presentations (see for instance [25]). The idea is to use elementary properties of graphs in the plane, in particular the Euler formula: $V - E + F = 2$. That is, given a non-empty finite connected planar graph, the sum of V, the number of vertices, and F, the number of components of the complement, minus E, the number of edges, is equal to 2.

Notice that in a van Kampen diagram over a finite presentation $\mathcal{P}$, when the boundaries of two compact faces have a common arc of intersection, the corresponding relators (read cyclically) contain a common subword. If these common subwords are always very short, then a face of the diagram which does not meet the boundary meets many other faces. Given an arc common to the boundaries of two faces, vertices of degree 2 can be suppressed and the label changed to the corresponding word in the generators. If all vertices of degree 2 are suppressed in this way, the planar graph now has all vertices of degree at least 3. A consequence of the Euler formula is that in a planar graph in which all vertices have degree at least three, there are at least two compact regions with at most 5 sides. That can be seen as follows:

If all vertices have degree at least 3, counting the edges at each vertex gives $2E \geq 3V$ and $V \leq 2E/3$ (as each vertex meets at least three edges, and each edge is counted twice).

If all faces except two (for instance the unbounded region and one compact region) have at least 6 sides (the other two having at least 1 side each), then $E \geq 6(F - 2)/2 + 2/2$ implies that $F \leq E/3 + 5/3$ (as each of $F - 2$ faces meets at least six edges, and each edge is counted twice).

Thus $V + F - E \leq E - E + 5/3 = 5/3$ and is therefore not 2.

Thus there are at least three regions with 5 or fewer edges.

In order to apply this to van Kampen diagrams, we need to slightly alter our view of the diagrams, and for this we need the following definitions.

Definition 3.1 (The small cancellation conditions). Let $\mathcal{P} = \langle \mathcal{A} \mid R \rangle$ be a finite presentation. Let R^C be the set of all reduced cyclic conjugates of elements of R and their inverses.

A non-trivial word $p = a_1 a_2 \ldots a_k \in F(\mathcal{A})$ is a *piece* with respect to R if there are two different relations $r_1, r_2 \in R^C$ such that $r_1 = ps_1$ and $r_2 = ps_2$ (and of course $s_1 \neq s_2$). That is, p occurs as a subword of two different relations in R^C. Notice that non-trivial subwords of pieces are also pieces.

The presentation satisfies the conditions:

- $C'(1/\lambda)$ for $\lambda \in \mathbb{R}^+$ if for all pieces p, if p is a subword of $r \in R^C$, then $\ell(p)/\ell(r) < 1/\lambda$.

- $C(k)$ if for all relations $r \in R^C$, it is not possible to write r as a product of fewer than k pieces.

Clearly the property $C'(1/\lambda)$ implies the property $C'(1/\lambda')$ if $\lambda \geq \lambda'$, and $C(k)$ implies $C(k')$ if $k' \leq k$.

The first non-trivial example to consider is the fundamental group of a genus two closed orientable surface, with the standard presentation $\langle a, b, c, d \mid aba^{-1}b^{-1}cdc^{-1}d^{-1} \rangle$. Here each generator is a piece, and no word of length 2 is a piece. Thus the presentation satisfies the conditions $C'(1/7)$ and $C(8)$.

Notice that for positive integers k, the metric condition $C'(1/k)$ implies the condition $C(k+1)$. For a presentation which satisfies the condition $C(k)$, in all *reduced* van Kampen diagrams, each compact face not meeting the boundary has at least k sides (when degree 2 vertices are suppressed). This is because when a long subword labels a segment separating two faces of a diagram, the relations must be equal in F (and so the same element of R^C), when read from the initial point of the common segment. It follows that the diagram is not reduced.

Lemma 3.2 (The small cancellation lemma : non-metric version). *Let Δ be a van Kampen diagram which is a topological disk, containing at least 2 regions, such that all internal regions have ≥ 6 sides, and all vertices have degree ≥ 3. For $i = 1, \ldots, 5$, let b_i be the number of regions meeting $\partial \Delta$ in exactly one connected segment (and possibly some vertices) having exactly i internal edges, forming a connected part of their (internal) boundary.*

Then $3b_1 + 2b_2 + b_3 \geq 6$.

One of the main applications of this important lemma is:

Theorem 3.3. *If a finite presentation $\mathcal{P}$ satisfies the condition $C'(1/6)$, then Dehn's algorithm solves the word problem.*

First we prove the theorem, supposing that the lemma has been proved.

Proof of the theorem. First note that $\mathcal{P}$ also satisfies the non-metric $C(7)$ (and so the $C(6)$) condition, and that the metric condition implies that the sum of the lengths of 3 pieces of a relator r is strictly less than half the length of r.

Let w be a cyclically reduced non-trivial word representing the identity element of the group, and let D be a diagram for w over $\mathcal{P}$. We show that w contains a subword which is more than half of a relator.

The form of the diagram is that of a collection of topological discs in the plane joined by trees, such that regarding the topological discs as fat vertices, the whole is a tree. As w is supposed to be cyclically reduced, there are no vertices of degree 1 to be suppressed (removing them and the edges meeting them would change w by a cyclic reduction). Either the diagram now consists of a single topological disc, or there are at least two extremal topological discs which intersect the rest of the diagram in just one vertex.

Case 1: D consists of a single topological disc. The lemma provides two faces F_1, F_2 meeting the boundary ∂D in segments of length $> \ell(F_1)/2$ and $> \ell(F_2)/2$. This means that the segment corresponding to $\partial D \cap \partial F_1$ is labelled by a subword u of some r, a cyclic conjugate of a relator (or its inverse) in R, of length greater than $\ell(r)/2$, as it is all of r except for at most 3 pieces. Thus $r = uv$ such that $\ell(u) > \ell(v)$, and so $u =_\Gamma v^{-1}$ and $w = w'uw''$ and $w =_\Gamma w'vw''$, and $\ell(w'vw'') < \ell(w'uw'')$.

Case 2: D contains more than one topological disc. There are at least two disc components which are extremal, i.e. which meet the rest of the diagram in a single vertex. In each of these components, at least one of the two boundary segments given by the lemma has an interior which does not contain this vertex. Thus the diagram contains two faces meeting the boundary in the manner claimed. $\qquad\square$

Proof of the lemma. The following proof is given by L.I. Greendlinger and M.D. Greendlinger in [16]. Let Δ be a reduced van Kampen diagram over the presentation $\mathcal{P}$ which is a topological disc, and suppose that Δ satisfies the $C(6)$ condition.

Create a diagram $\mathcal{G}$ on S^2 by doubling Δ along $\partial\Delta$, after a twist by less than half the length of the shortest edge on $\partial\Delta$. This gives a graph on S^2.

The twist creates a new vertex in each edge of $\partial\Delta$. Each region in Δ contributing to b_i has $i+1$ sides in its boundary, and the region intersects the interior of Δ in a consecutive series of i of these edges (and possibly some other vertices). In this way, every region of Δ contributing to b_1, b_2, b_3 gives 2 regions in $\mathcal{G}$, each with 3,4, and 5 sides respectively.

All other bounded regions of Δ give 2 regions in $\mathcal{G}$ each with at least 6 sides (including regions meeting the boundary in more than one connected segment).

Let V, E, F denote the numbers of vertices, edges and regions of $\mathcal{G}$. Then

$$E \geq (6(F - 2b_1 - 2b_2 - 2b_3) + 3.2b_1 + 4.2b_2 + 5.2b_3)/2$$
$$\implies E \geq 3F - 3b_1 - 2b_2 - b_3 \implies F \leq E/3 + (3b_1 + 2b_2 + b_3)/3.$$

Also $E \geq 3V/2$ implies $V \leq 2E/3$, so

$$2 = V - E + F \leq 2E/3 - E + E/3 + (3b_1 + 2b_2 + b_3)/3$$

which in turn implies that $3b_1 + 2b_2 + b_3 \geq 6$.   □

It is not hard to give a second condition on the combinatorics of a presentation which guarantees that internal vertices have degree at least 4:

Definition 3.4. Let $\mathcal{P} = \langle \mathcal{A} \mid R \rangle$ be a finite presentation. Let R^C be the set of all reduced cyclic conjugates of elements of R and their inverses. The presentation satisfies the condition $T(q)$ if :
for all $3 \leq k < q$, for all $r_1, r_2, \ldots, r_k \in R^C$, and for all pieces $p_1, \ldots, p_k$, if $r_1 = p_1 r_1' p_2^{-1}$, $r_2 = p_2 r_2' p_3^{-1}, \ldots$, and $r_k = p_k r_k' p_1^{-1}$, then for some $i \mod k$, it is the case that $r_i = r_{i+1}^{-1}$.

This complicated-looking definition just states that in a *reduced* diagram, all internal vertices have degree at least q.

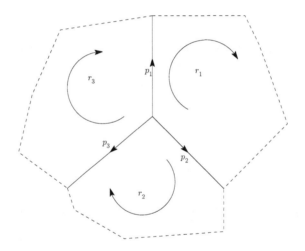

Figure 3.1: The $T(q)$ condition with $k = 3$.

With this definition, the same methods as above can be used to show:

Lemma 3.5 (The small cancellation lemma $C(4) - T(4)$ version). *Let $\mathcal{P}$ be a finite presentation satisfying the conditions $C(4)-T(4)$, and D be a reduced van Kampen diagram over $\mathcal{P}$ which is a topological disc containing more than one face.*

Then D contains two bounded faces $\partial F_1, \partial F_2$, such that for $i = 1, 2$, $\partial F_i \cap \partial D$ contains a connected segment containing all of ∂F_i except for at most 2 pieces.

Theorem 3.6. *A presentation which satisfies $C(6)$ or $C(4) - T(4)$, satisfies a quadratic isoperimetric inequality.*

Proof. We consider a van Kampen diagram Δ satisfying the $C(6)$ condition, where vertices of degree 2 have been suppressed.

We first consider the case when Δ is a topological disk.

Note that if F has a non-boundary edge, both of its endpoints lying on the boundary, splitting Δ into two subdiagrams Δ_1, Δ_2, each with $\delta_i \geq 4$ vertices of degree at least 3 on its boundary, then $\delta_1 + \delta_2 + 2 \geq \delta$. The induction hypothesis, that the number of faces is bounded by the square of the number of vertices on the boundary of a topological disc diagram (when vertices of degree 2 are suppressed) tells us that $F = F_1 + F_2 \leq {\delta_i}^2 + {\delta_2}^2 \leq (\delta_1 + \delta_2 - 2)^2$ if $\delta_1, \delta_2 \geq 4$.

We now suppose that there are no such edges with both endpoints on the boundary. Thus there any region of Δ meeting $\partial\Delta$ in at least two segments has at least 6 sides in Δ.

We use the notation of 3.2, and in addition we denote by b_5' the sum of $\sum_{i>4} b_i$ and the number of regions of Δ meeting $\partial\Delta$ in at least one segment and having at least 5 sides, and we denote by V' number of vertices in Δ having degree at least 4.

Notice that the regions contributing to b_4 give rise to regions in $\mathcal{G}$ with 6 sides, while those contributing to b_5' give rise to regions with at least 7 sides.

Counting the edges in $\mathcal{G}$ via the faces, we get

$$E \geq 3(F - 2(b_1 + b_2 + b_3 + b_4 - b_5')) + 3b_1 + 4b_2 + 5b_3 + 6b_4 + 7b_5'$$
$$\geq 3F - 3b_1 - 2b_2 - b_3 + b_5'$$
$$\implies F \leq E/3 - (3b_1 + 2b_2 + b_3 - b_5')/3.$$

Estimating the number of edges of $\mathcal{G}$ via the vertices, we get

$$E \geq (3(V - 2V') + 4.2V')/2 = 3V/2 + V'$$
$$\implies V \leq (2E - 2V')/3.$$

Applying these inequalities to the Euler characteristic of the graph $\mathcal{G}$ on the sphere, we get

$$2 = V - E + F \leq (2E - 2V')/3 - E + E/3 - (3b_1 + 2b_2 + b_3 - b_5')/3$$
$$\implies 6 \leq (3b_1 + 2b_2 + b_3) - (2V' + b_5').$$

It follows that there are at least two more regions meeting the boundary of Δ contributing to $(3b_1 + 2b_2 + b_3)$ than there are regions and vertices contributing to $(2V' + b_5')$. If, between each pair of vertices contributing to $3b_1 + 2b_2 + b_3$, there is a boundary vertex contributing to $(2V' + b_5')$, then we would have $2V' + b_5' \geq V' \geq 3b_1 + 2b_2 + b_3$. From this we deduce that there is a segment s on $\partial\Delta$ meeting faces $F_1, F_2, \ldots, F_k$ in subsegments $s_1, s_2, \ldots, s_k$ such that

- $\partial F_i \cap S = s_i$;

- $s_i \cap s_{i+1} = v_i$, a vertex in $s \in \partial\Delta$ of degree 3;

- F_1 and F_k contribute to $3b_1 + 2b_2 + b_3$: have boundary consisting of one edge in $s \subset \partial\Delta$ together with a sequence of at most 3 interior edges;

- for $i = 2, \ldots, k-1$, each region F_i contributes to b_4, and it has a boundary consisting of one edge in $s \subset \partial\Delta$ together with a sequence of at most 3 interior edges;

- for $i = 2, \ldots, k-1$, each intersection $F_i \cap F_{i+1}$ contains an edge which meets the vertex v_i, the other end being the vertex u_i.

Also, if the first vertex v_0 has degree 4, then it will contribute to $2V'$, though it may be counted twice if it is also the final vertex of another chain. In any case, we may assume that v_0 and v_k, the final vertex of F_k are of degree 3.

We show that for this (topological disk diagram) Δ, the number of faces is bounded above by the square of δ, the number of vertices in $\partial\Delta$.

Consider the edge f_i of the chain: the vertices of this edge are $v_i = f_i \cap s$ and $u_i \notin \partial\Delta$. Suppose that one of the other edges at u_i, say e, has an endpoint on $\partial\Delta$ (and so u_i has degree at least 4). Then cutting Δ along the edges f_i and e splits Δ into two diagrams as before, each of which has at least 4 vertices on its boundary, and as before the induction hypothesis applies to show that $F \leq \delta^2$.

We can now suppose that none of the vertices u_i associated to the chain s is joined to $\partial\Delta$ (other than by f_i). If in each such chain there were a vertex u_i of degree at least 4, then again $V' > 3b_1 + 2b_2 + b_3$. Thus we can suppose that each vertex u_i has degree 3.

Removing the faces $F_1, \ldots, F_k$ and the edges $s_1, \ldots s_k, f_1, \ldots, f_{k-1}$, from Δ gives a diagram Δ' which is a topological disc with k fewer faces. The vertices v_0 and v_k have degree 2, so that after their suppression, and the suppression of the vertices u_i, all now of degree 2, the number δ' of vertices in boundary of Δ' is at most $\delta - 2$. Thus $F = F' + k \leq (\delta - 2)^2 + k \leq \delta^2 - 2\delta + 4 + k$. But $k \leq \delta$ and $\delta \geq 4$ so that $F \leq \delta^2$.

Now that the quadratic inequality has been established for topological disk diagrams without vertices of degree 2, it suffices to notice that a general van Kampen diagram without vertices of degree 2 satisfies the same inequality, as such a diagram is made up of several disc diagrams, the area is the sum of the areas of the disc components, and the length of the boundary is at least the sum of the boundary lengths.

Finally, to recover the actual isoperimetric inequality for the presentation, the vertices of degree 2 must be re-instated. But on any edge in the boundary of a region corresponding to the relation r, there at most $\max \ell(r)$ vertices.

Thus $Area(w) \leq \max \ell(r)\ell(w)^2$.

The result for $C(4) - T(4)$ presentations follows in the same way. $\qquad \square$

The same sort of proof can be used to solve conjugacy problems, using diagrams on an annulus rather than a disk. One application of this is Weinbaum's solution [30] of the conjugacy problem for alternating knot groups (see also [20, Chapter V]). The first step in the proof is to show that a prime tame alternating knot k in S^3 gives rise to a presentation for $\pi_1(S^3 - k) * \mathbb{Z}$ which satisfies the conditions $C(4) - T(4)$.

Chapter 4

Isoperimetric Inequalities and Quasi-Isometries

As we have stated earlier, solving the word problem for a presentation $\mathcal{P} = \langle \mathcal{A} \mid R \rangle$ involves showing that an expression for w as a product of conjugates of relators $w = \prod_{i=1}^{M} p_i r_i^{\epsilon_i} p_i^{-1}$ can be found, with $p_i \in F(\mathcal{A})$, $r_i \in R$, and $\epsilon_i = \pm 1$. The smallest such number M is called the area of w. The function $\delta_{\mathcal{P}} : \mathbb{N} \to \mathbb{N} :$ $\delta_{\mathcal{P}}(n) = \max\{Area(w) \mid w \in \langle\langle R \rangle\rangle, \ell(w) \leq n\}$ is called the Dehn function of the presentation. An *isoperimetric inequality* for the presentation is a function $f : \mathbb{N} \to \mathbb{R}$ such that for all $n \in \mathbb{N}$, $\delta_{\mathcal{P}}(n) \leq f(n)$.

Geometric group theory is concerned with properties of groups which are invariant under quasi-isometries (this is almost a definition of the theory):

Definition 4.1. Let $\lambda \geq 1$ and $\epsilon \geq 0$ be constants, and let X, Y be metric spaces. A map $f : X \to Y$ is a (λ, ϵ)-*quasi-isometry*, or a (λ, ϵ)*quasi-isometric embedding*, if for every pair of points $x, x' \in X$ we have

$$\frac{1}{\lambda} d_X(x, x') - \epsilon \leq d_Y(f(x), f(x')) \leq \lambda d_X(x, x') + \epsilon.$$

If there is such a (λ, ϵ)-quasi-isometry and a constant C such that for all $y \in Y$, there is a point $x(y) \in X$ such that $d_Y(f(x(y)), y) < C$ (i.e. if f is 'almost surjective') then we say that X and Y are quasi-isometric. The function $Y \to X$ given by $y \to x(y)$ is a *quasi-inverse* of the function f. In general a map $g : Y \to X$ is a quasi-inverse of the map f if there is a constant C such that $\forall x \in X, d_X(x, g \cdot f(x)) < C$ and $\forall y \in Y, d_Y(y, f \cdot g(y)) < C$.

Warning: care must be taken here as a quasi-isometry $f : X \to Y$ in general is far from being a surjection: for instance the inclusion of $\mathbb{R}$ in $\mathbb{R} \times \mathbb{R}$ as a factor is a quasi-isometry (a quasi-isometric embedding).

The standard examples are:

Bounded spaces are quasi-isometric to a point.

The inclusion $\mathbb{Z} \hookrightarrow \mathbb{R}$ is a quasi-isometry, and in fact is almost onto, with the function "integer part" providing a quasi-inverse.

The inclusion $\mathbb{Z} \times \mathbb{Z} \hookrightarrow \mathbb{R} \times \mathbb{R}$, can be viewed as a quasi-isometric embedding of a Cayley graph of the fundamental group of a torus in the universal covering space, and this inclusion is almost onto.

In the same way, the tessellation of the hyperbolic plane $\mathbb{H}^2$ by regular octagons with corner angle $\pi/4$ describes the universal covering space of a surface S_2 of genus 2. The dual graph can be viewed as a quasi-isometric embedding of the Cayley graph of the fundamental group $\pi_1(S_2)$ in $\mathbb{H}^2$. This is a special case of the result due to Švarc [28] and to Milnor [23] that the universal covering space of a closed compact Riemannian manifold is quasi-isometric to the Cayley graph of the fundamental group. We shall prove a more general form of the statement, following Bridson and Haefliger's approach [6, p.140]. In order to state the result in a little more generality, we introduce the following definitions: a metric space (X, d) is *proper* if closed balls $\overline{B}_r(x)$ are compact, and is a *length space* if for all points $x, y \in X$, $d(x, y) = \inf \ell(\gamma)$, where the infimum is taken over all rectifiable curves in X from x to y. Recall that a curve $\gamma : [0, 1] \to X$ is rectifiable of length L if $L = \sup \sum_{i=0}^{M-1} d_X(\gamma(t_i, t_{i+1})$ is finite, where the supremum is taken over all subdivisions $0 = t_0 < t_1 < \cdots < t_M = 1$ of the interval $[0, 1]$.

A group Γ acts *properly* on a metric space X if for all compact $K \subset X$, the set $\{g \in \Gamma \mid K \cap g \cdot K \neq \emptyset\}$ is finite.

Proposition 4.2. *Let (X, d_X) be a proper length space, and suppose that Γ acts properly and cocompactly on X. Then*

- *Γ is finitely generated;*

- *for any point $z \in X$, there is a finite generating set $\mathcal{A}$ for Γ such that the map $Cay^1(\Gamma, \mathcal{A}) \to X$ defined on the vertices by $g \to g \cdot z$ is a quasi-isometry on the 0-skeleton.*

Remarks. i) We shall see that the existence of a quasi-isometry between the Cayley graph and the space X is independent of the choice of finite generating set.

ii) The quasi-isometry can be extended over the edges by defining arbitrarily the images of the edges originating at the identity element, and then defining the images of the other edges using the Γ action (or alternatively by mapping each edge to the image of its initial vertex).

iii) Consider the standard hyperbolic structure on the surface S of genus two. That is, consider S as the quotient of the hyperbolic plane $\mathbb{H}^2$ by the action of a discrete group of hyperbolic isometries with fundamental domain a regular octagon with corner angle $\pi/4$. The plane $\mathbb{H}^2$ is a length space: in fact the surface S obtained in this way is a Riemannian manifold and between any pair of points there is a unique geodesic whose length is the distance between the points.

iv) Generalising the surface example, if M is a closed compact manifold where the distance between points is the infimum of the lengths of rectifiable curves between them (for instance when M has a Riemannian metric), then the universal covering space (with the lift of the distance from M) provides the length space X upon which $\pi_1(M)$ acts freely.

Proof. Let $C \subset X$ be a compact fundamental domain for the action of Γ, i.e. such that $\bigcup_{g \in \Gamma} g \cdot C = X$. Choose $z \in X$ and $r \geq 1$ such that $C \subset B_r(z)$.

We shall show that the finite set $\mathcal{A} = \{g \in \Gamma \mid \overline{B}_{3r}(z) \cap g \cdot \overline{B}_{3r}(z) \neq \emptyset\}$ generates Γ. Note that $y \in g \cdot \overline{B}_{3r}(z) \cap \overline{B}_{3r}(z) \implies g^{-1} \cdot y \in \overline{B}_{3r}(z) \cap g^{-1} \cdot \overline{B}_{3r}(z)$, so that $\mathcal{A}^{-1} = \mathcal{A}$.

To establish the quasi-isometry, we show:

1) $\exists \lambda_1 > 0$ and $\epsilon_1 \geq 0$ s.t. $d_{\mathcal{A}}(1, g) \leq \lambda_1 d_X(z, g \cdot z) + \epsilon_1$.

2) $\exists \lambda_2$ such that $\forall g, g' \in \Gamma$, $d_X(g \cdot z, g' \cdot z) \leq \lambda_2 d_{\mathcal{A}}(g, g')$.

We begin by proving 1), and that $\mathcal{A}$ is a finite generating set.

As X is a length space, there is a rectifiable path $\gamma : [0, 1] \to X$ such that $\ell(\gamma) \leq d_X(z, g \cdot z) + 1$, and we can choose a subdivision $0 = t_0 < t_1 < \cdots < t_M = 1$ such that $z = \gamma(0)$, $g \cdot z = \gamma(1)$, and $d_X(\gamma(t_i), \gamma(t_i + 1)) = r$ for $i < M - 1$ and $d_X(\gamma(t_{M-1}), \gamma(t_M)) \leq r$. Thus $Mr \leq \ell(\gamma) \leq (M + 1)r$, and $Mr \leq d_X(z, g \cdot z) + 1 \implies M \leq d_X(z, g \cdot z)/r + 1$.

Put $g_0 = 1$, $g_M = g$, and for each $1 \leq i \leq M - 1$, choose $g_i \in \Gamma$ such that $d_X(g_i \cdot z, \gamma(t_i)) \leq r$. This is possible as $\bigcup_g g \cdot B_r(z) = X$.

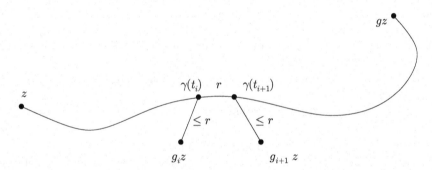

Figure 4.1: Proving finite generation

It follows that $d_X(g_i \cdot z, g_{i+1} \cdot z) \leq 3r \implies d_X(z, g_i^{-1} g_{i+1} \cdot z) \leq 3r$, and so $g_i^{-1} g_{i+1} \in \mathcal{A}$. But $g = g_M = 1.g_1.(g_1^{-1} g_2). \ldots . (g_{M-1}^{-1} g_M)$, and it follows that $\mathcal{A}$ generates Γ, and $d_{\mathcal{A}}(1, g) \leq M$. Moreover $d_X(z, g \cdot z) \geq Mr - 1 \geq d_{\mathcal{A}}(1, g)r - 1 \implies d_{\mathcal{A}}(1, g) \leq (1/r)d_X(z, g \cdot z) + (1/r)$.

To conclude, it suffices to take $\lambda_1 = \epsilon_1 = 1/r$.

We now proceed to the proof of 2).

As $d_X(g \cdot z, g' \cdot z) = d_X(z, g^{-1} g' \cdot z)$, it suffices to consider the case $g' = 1$.

We know from 1) that $\mathcal{A}$ is a finite generating set. In terms of this generating set, there is a shortest word $a_1 a_2 \ldots a_n$ representing g in Γ; let $g_i = a_1 \ldots a_i$, for $i = 1, \ldots, n - 1$.

Then $d_X(z, g \cdot z) \leq d_X(z, g_1 \cdot z) + d_X(g_1 \cdot z, g_2 \cdot z) + \cdots + d_X(g_{n-1} \cdot z, g \cdot z) \leq \lambda_2 n$ where $\lambda_2 = \max_{a \in \mathcal{A}} d_X(z, a \cdot z)$. But $n = d_{\mathcal{A}}(1, g)$, and 2) holds.

Summing up, we see from 1) that $d_{\mathcal{A}}(1, g) \leq (1/r)d_X(z, g \cdot z) + (1/r)$, and from 2) that $1/\lambda_1 d_X(z, g \cdot z) \leq d_{\mathcal{A}}(1, g)$. The inclusion of $Cay^1(\Gamma, \mathcal{A}) \to X$ is thus a $(\max\{1, \lambda_1, \lambda_2\}, r)$-quasi-isometry on the 0-skeleton.   □

The first (and perhaps the main) example of interest here concerns the maps obtained between Cayley graphs by changing the (finite) generating set of a finitely generated group. The above proof for good group actions apparently depends on the generating set used. The whole point of studying the geometry of the Cayley graph is to obtain properties of the *group* from the graph, so these properties must be invariant under change of finite generating set.

Proposition 4.3. *Let $\mathcal{A}$ and $\mathcal{B}$ be two finite generating sets for the group Γ. Then the Cayley graphs $Cay^1(\Gamma, \mathcal{A})$ and $Cay^1(\Gamma, \mathcal{B})$ are quasi-isometric via a map which is the identity on the vertices.*

Proof. Let $\mathcal{A} = \{x_1, \ldots, x_p\}$, and $\mathcal{B} = \{y_1, \ldots, y_q\}$. For each x_i (respectively y_j), there is a word $u_i(\mathcal{B}) \in F(\mathcal{B})$ (resp. $v_j(\mathcal{A}) \in F(\mathcal{A})$) representing the same element as x_i (resp. y_j) in the group Γ.

Consider two elements $g, g'' \in \Gamma$, and suppose that $d_{\mathcal{A}}(g, g') = k$. Then there is a word $w(X) = x_{i_1}^{\epsilon_1} \ldots x_{i_k}^{\epsilon_k}$ in $F(\mathcal{A})$, such that $gw =_\Gamma g'$. Translating into the $\mathcal{B}$ generating set, gives the word $w(\mathcal{B}) = u_{i_1}^{\epsilon_1}(\mathcal{B}) \ldots u_{i_k}^{\epsilon_k}(\mathcal{B})$ such that $w(\mathcal{B}) =_\Gamma w(\mathcal{A})$, and the word $w(\mathcal{B})$ labels a path in the Cayley graph $Cay^1(\Gamma, \mathcal{B})$ from the vertex labelled g to the vertex labelled g'.

Let $K = \max\{\ell_{\mathcal{B}}(u_i(\mathcal{B})) \mid i = 1, \ldots, p\}$; then $\ell_{\mathcal{B}}(w(\mathcal{B})) \leq Kk$ and so $d_{\mathcal{B}}(g, g') \leq Kd_{\mathcal{A}}(g, g')$.

In the same way, setting $K' = \max\{\ell_{\mathcal{A}}(v_j(\mathcal{A})) \mid j = 1, \ldots, q\}$, we see that $d_{\mathcal{A}}(g, g') \leq K'd_{\mathcal{B}}(g, g')$.

Thus $(1/K')d_{\mathcal{A}}(g, g') \leq d_{\mathcal{B}}(g, g') \leq Kd_{\mathcal{A}}(g, g')$.

In this way the identity map on the vertices is a quasi-isometry. To define a quasi-isometry on the edges, it suffices to extend to the identity map on the vertices to maps $\phi : Cay^1(\Gamma, \mathcal{A}) \to Cay^1(\Gamma, \mathcal{B})$ and $\psi : Cay^1(\Gamma, \mathcal{B}) \to Cay^1(\Gamma, \mathcal{A})$ taking the whole interior of each edge to either of its endpoints. Let t, t' be points in the interiors of two edges in $Cay^1(\Gamma, \mathcal{A})$, and consider $\phi(t), \phi(t')$ as elements of Γ, which can be viewed as a subset of $Cay^1(\Gamma, \mathcal{A})$ or of $Cay^1(\Gamma, \mathcal{B})$. Then $d_{\mathcal{A}}(t, t') \leq 1$, so that $d_{\mathcal{B}}(\phi(t), \phi(t')) \leq Kd_{\mathcal{A}}(t, t') + 2K$.

Also clearly $d_{\mathcal{A}}(\psi \cdot \phi(t), t) \leq 1$ and $d_{\mathcal{B}}(\phi \cdot \psi(s), s) \leq 1$ for all $s \in Cay^1(\Gamma, \mathcal{B})$. The maps ϕ and ψ are quasi-inverses, and the proof is complete.   □

Definition 4.4. Two finitely generated groups are said to be quasi-isometric if they have quasi-isometric Cayley graphs with respect to some (and hence all by 4.3) finite generating sets. A group-theoretic property is said to be *geometric* if it is invariant under quasi-isometry of groups.

The first essential property is that of being finitely presented ([1]):

Proposition 4.5. *Let $\mathcal{A}$ and $\mathcal{B}$ be finite generating sets for the groups Γ and Γ'. If $Cay^1(\Gamma, \mathcal{A})$ and $Cay^1(\Gamma', \mathcal{B})$ are quasi-isometric and Γ is finitely presentable, then Γ' is also finitely presentable.*

We shall in fact show something much stronger from which this proposition can be deduced: we shall show that quasi-isometric groups have comparable Dehn functions.

Definition 4.6. We say that two functions $f, g : \mathbb{N} \to \mathbb{R}$ are equivalent if there is a positive constant A such that for all $n \in \mathbb{N}$, $f(n) \leq Ag(An + A) + An + A$ and $g(n) \leq Af(An + A) + An + A$.

Notice that with this definition, linear functions are equivalent to constant functions (even to the zero function) and that all polynomials of degree $d > 1$ form an equivalence class, and all exponential functions form an equivalence class. It thus makes sense to talk about groups satisfying a linear, quadratic or exponential isoperimetric inequality, once we have shown the following result:

Theorem 4.7 ([1]). *Let $\mathcal{P} = \langle \mathcal{A} \mid R \rangle$ be a finite presentation of the group Γ. Let $\mathcal{B}$ be a finite generating set for the group Γ' such that $Cay^1(\Gamma, \mathcal{A})$ and $Cay^1(\Gamma', \mathcal{B})$ are quasi-isometric. Then there is a finite set of relators S for Γ' such that $\mathcal{Q} = \langle \mathcal{B} \mid S \rangle$ is a finite presentation for Γ', and the Dehn functions for the presentations $\mathcal{P}$ and $\mathcal{Q}$ are equivalent.*

An immediate consequence of this is that "having solvable word problem" is a geometric property, and in particular:

Corollary 4.8. *If $\mathcal{P}$ and $\mathcal{Q}$ are finite presentations of the group Γ and the word problem is solvable for $\mathcal{P}$, then the word problem is solvable for $\mathcal{Q}$.*

Proof of the theorem. Let $\phi : Cay^1(\Gamma, \mathcal{A}) \to Cay^1(\Gamma', \mathcal{B})$ be a quasi-isometry and let $\psi : Cay^1(\Gamma', \mathcal{B}) \to Cay^1(\Gamma, \mathcal{A})$ be a quasi-inverse, so that for all vertices $g \in Cay^1(\Gamma', \mathcal{B})$, $d_\mathcal{B}(\phi \cdot \psi(g), g) \leq C$. Up to changing the quasi-isometry constants, we can suppose that vertices are sent to vertices, and that the vertices corresponding to the identity elements in each group are sent to each other.

Let $w = y_{j_1} \ldots y_{j_k} \in F(\mathcal{B})$ be a word labelling a closed loop in $Cay^1(\Gamma', \mathcal{B})$ based at the vertex 1. For each initial segment $w_m = y_{j_1} \ldots y_{j_m}$, for $m = 1, \ldots, k$, let $v_m \in Cay^1(\Gamma', \mathcal{B})$ be the vertex represented by the word w_m. Consider the sequence of points $u_0 = 1, u_1 = \psi(v_1), \ldots, u_k = \psi(v_k) = 1 \in Cay^1(\Gamma, \mathcal{A})$. As ψ is a (λ, ϵ)-quasi-isometry, and $d_\mathcal{B}(v_i, v_{i+1}) = 1$, we have $d_\mathcal{A}(u_i, u_{i+1}) \leq \lambda + \epsilon$.

There is therefore for each $i = 0, \ldots, k$ a word $\alpha_i \in F(\mathcal{A})$ of length at most $\lambda + \epsilon$, such that there is a path in $Cay^1(\Gamma, \mathcal{A})$ labelled α_i from the vertex u_i to the vertex u_{i+1}. The product $w' = \alpha_0 \alpha_1 \ldots \alpha_k$ labels a loop in $Cay^1(\Gamma, \mathcal{A})$. There is therefore a van Kampen diagram D for w' over $\mathcal{P}$. The 1-skeleton of this diagram maps into the Cayley graph $Cay^1(\Gamma, \mathcal{A})$. Applying $\phi : D^{(1)} \to Cay^1(\Gamma', \mathcal{B})$, corresponds to relabelling the diagram D, to get a diagram D' where each edge which was labelled by a generator in $\mathcal{A}$, is now labelled by a word in $F(\mathcal{B})$ of length at most $\lambda + \epsilon$. Each compact face of D was labelled by a word $r \in R$, and is now labelled by a word of length at most $\lambda\ell(r) + \epsilon$ in $F(\mathcal{B})$. The boundary of D, which was labelled by w' is now labelled by $w'' = \phi(\alpha_0)\phi(\alpha_1)\ldots\phi(\alpha_k)$. Considering the path labelled w'' based at the vertex 1, the vertex reached by the initial segment $\phi(\alpha_0)\phi(\alpha_1)\ldots\phi(\alpha_m)$ is the vertex $\phi \cdot \psi(v_m)$, and so is at distance at most C from the vertex v_m.

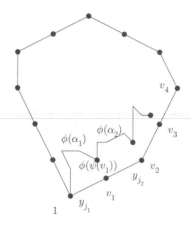

Figure 4.2: Part of the paths w and $\phi(\psi(w))$ in $Cay^1(\Gamma', \mathcal{B})$

It follows that we can construct a van Kampen diagram for w over $\mathcal{P}'$ by adding to the diagram D' (the relabelled diagram D) for each $m = 0, \ldots, k$ a diagram for the word $\phi(\alpha_m)y_{j_m}^{-1}h_m$, where h_m is a word of length at most C. In this way we obtain a van Kampen diagram for w of area $const.area_\mathcal{P}(\psi(w)) + const.\ell(w)$, and w is in the normal closure of set of relations in $F(\mathcal{B})$ of length at most $\max(C+1+\lambda+\epsilon, (\lambda+\epsilon)\max_{r\in R}\{\ell(r)\})$. As $\ell(\psi(w)) \leq const.\ell(w)$, this also shows an isoperimetric inequality for Γ gives an equivalent isoperimetric inequality for Γ'. □

The special class of hyperbolic groups is the class of all finitely presented groups satisfying a linear isoperimetric inequality. An alternative definition is via the definition of "thin triangles".

Definition 4.9. Let (X, d) be a geodesic metric space (length space) — i.e. where, between any two points there is a path (a "geodesic") whose length is equal to the distance between the points.

For any three points x, y, z there is a "geodesic triangle" $\Delta(x, y, z)$ formed by taking a geodesic between each pair of points. (There may be many such geodesics.)

Because of the triangle inequality, for each such triangle, there is a Euclidean triangle $\Delta(x', y', z')$ with the same side lengths. The Euclidean triangle maps onto a tripod $Y(x'', y'', z'')$, by collapsing the inscribed circle onto a point.

Let T_Δ be the composite map from the (edges of the) triangle $\Delta \to Y$. For a positive real number δ, we say that the triangle Δ is δ-*thin* if $\forall p \in Y$ the diameter $Diam(T^{-1}(p)) \le \delta$.

The space X is said to be δ-hyperbolic if every geodesic triangle is δ-thin.

A finitely generated group Γ is said to be *hyperbolic* if it has a finite generating set $\mathcal{A}$ such that the Cayley graph $Cay^1(\Gamma, \mathcal{A})$ is δ-hyperbolic for some positive δ.

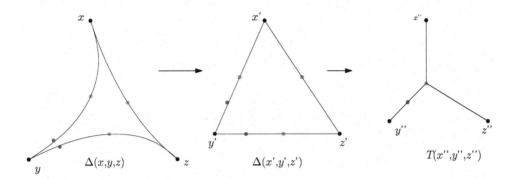

Sometimes δ-thin triangles are called *uniformly* δ-thin triangles (see [2]). An alternative definition is to say that a space is δ-hyperbolic space if for any geodesic triangle, each side lies in a δ-neighbourhood of the union of the other two sides. It is not to hard to show that the two definitions are equivalent (though it may be necessary to change the value of the constant δ). In order to show that all Cayley graphs (with respect to any finite generating sets) of a hyperbolic group are hyperbolic one must show that quasi-geodesic triangles (i.e. images of geodesic triangles under a quasi-isometry) are also thin, or some other equivalent result.

Before showing that hyperbolic groups satisfy a linear isoperimetric inequality, we shall first show that they satisfy a quadratic isoperimetric inequality, as this proof is simple and illustrates well the ideas of geometric group theory. A proof that they satisfy an isoperimetric inequality of type $n \log n$ is given in Section 6.2. We finally give in Proposition 4.11 a proof that they satisfy a linear isoperimetric inequality which is due to Noel Brady. It would suffice to show that geodesics in a hyperbolic group fellow travel, and then use Theorem 5.2.2 of Riley's notes to oobtain a linear isoperimetric inequality. In fact this same theorem states that a subquadratic isopermietric inequality implies a linear one (though this does use asymptotic cones!).

Proposition 4.10. *Let $\mathcal{A}$ be a finite generating set for the group Γ such that the Cayley graph $Cay^1(\Gamma, \mathcal{A})$ is δ-hyperbolic. Then Γ is finitely presentable and satisfies a quadratic isoperimetric inequality.*

Proof. We shall suppose that no generator in $\mathcal{A}$ is trivial in Γ.

Let $w = a_1 \ldots a_n \in F(\mathcal{A})$ be a word which represents the identity element of Γ. Then the word labels a closed loop based at the identity vertex of $Cay^1(\Gamma, \mathcal{A})$.

Let γ_i be a shortest word in $F(\mathcal{A})$ representing the element $a_1 \ldots a_i$ in Γ. Then the paths based at 1 in $Cay^1(\Gamma, \mathcal{A})$ with labels γ_i and γ_{i+1} form a geodesic triangle, together with the edge labelled a_{i+1} based at the vertex $a_1 \ldots a_i$ in $Cay^1(\Gamma, \mathcal{A})$. The fact that geodesic triangles are δ-thin means that this triangle can be decomposed as a collection of rectangles each of perimeter at most $2\delta + 2$ (the last is perhaps a triangle of perimeter at most $2\delta + 1$). There are at most $\max\{\ell(\gamma_i), \ell(\gamma_{i+1})\} \le n/2$ of these rectangles.

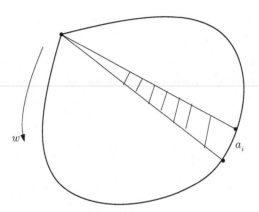

It follows that the set of relations $R = \{r \in F(\mathcal{A}) \mid \ell(r) \le 2\delta + 2, \ r =_\Gamma 1\}$ gives a finite presentation for Γ, and in terms of these relations, $area(w) \le n^2/2$. $\quad\square$

After this first simple proof, let us now show that in fact the group satisfies a linear isoperimetric inequality.

Proposition 4.11. *Let $\mathcal{A}$ be a finite generating set for the group Γ. If all geodesic triangles in $Cay^1(\Gamma, \mathcal{A})$ are δ-thin, then Γ is finitely presentable, has a Dehn presentation and satisfies a linear isoperimetric inequality*

Proof. The method used here (due to Noel Brady) is to show that "local geodesics" are like geodesics, and so Dehn's algorithm, with an appropriate set of relators solves the word problem for Γ.

For $k > 0$, a word $w \in F(\mathcal{A})$ in $Cay^1(\Gamma, \mathcal{A})$ is a k *local geodesic* if all subwords of w of length at most k are geodesic. We shall show that $2\delta + 2$ local geodesics do not label loops:

If a word is not a $2\delta + 2$ local geodesic, then it contains a subword v of length at most $2\delta + 2$ such that there is a shorter word u such that $v =_\Gamma u$. Take as relators $R = \{r \in F(\mathcal{A}) \mid \ell(r) \le 4\delta + 3, r =_\Gamma 1\}$. Dehn's algorithm, using this set of relators, can thus be used to convert any word into a $2\delta + 2$ local geodesic word representing the same element of the group (and in fact this can be done in time depending linearly on the length of the original word [2, 2.18] — it can even be done in *real time*, as has been shown by Holt and Röver [18]).

Claim: if $w =_\Gamma 1$ (and so w labels a loop in $Cay^1(\Gamma, \mathcal{A})$, then w is not a $2\delta + 2$ local geodesic (i.e. it contains a subsegment of length at most $2\delta + 2$ which is not geodesic).

Proof of the claim. We argue by contradiction: let w be a non-empty word in $F(\mathcal{A})$ which represents the trivial element of Γ (labels a loop in $Cay^1(\Gamma, \mathcal{A})$) and suppose that w is a $2\delta + 2$ local geodesic: the length of w is then at least $4\delta + 4$.

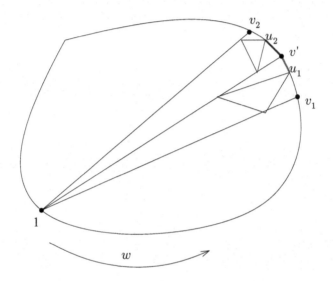

Let γ be a loop in $Cay^1(\Gamma, \mathcal{A})$ based at the vertex 1 and labelled by the $2\delta + 2$ local geodesic word w. Let v' be a vertex on γ furthest from the base point 1. This point is at distance at least $2\delta + 1$ from 1, else w is trivial (the letter in the $2\delta + 2$ position of w labels an edge ending at distance $2\delta + 2$ from 1). Let v_1, v_2 be the vertices on γ before and after v' at distance $2\delta + 1$ from v'.

Consider a geodesic triangle Δ_1 (resp. Δ_2) with vertices $1, v', v_1$ (resp. $1, v', v_2$) with one side the segment γ_1 (resp. γ_2) of γ between v' and v_1 (resp. v_2) of length $2\delta + 1$. Let u_1 (resp. u_2) be the point on γ_1 (resp. γ_2) mapping to the central point of the tripod under the tripod map T_{Δ_1} (resp. T_{Δ_2}). Thus $|d_X(1, v_1) - d_X(1, v')| = |d_X(v_1, u_1) - d_X(u_1, v')|$.

If $d_X(u_1, v') < \delta + 1$, then $d_X(v_1, u_1) \geq 2\delta + 1 - (\delta + 1) \geq d_X(u_1, v')$ and so $d_X(v_1, 1) > d_X(v', 1)$ contradicting the choice of v'.

In the same way, we see that $d_X(u_2, v') \geq \delta + 1$. It follows that the points u_1', u_2' at distance $\delta + 1$ before and after v' on the segment of γ containing v' lie between u_1 and u_2. But u_1' lies at distance $\delta + 1$ from v' on the geodesic from 1 to v', as does u_2', and so $d_X(u_1', u_2') \leq 2\delta$, contradicting the fact that w is assumed to be a $2\delta + 1$ local geodesic. $\square$

In this proof, if the path γ is not assumed to be a loop, what is proved is that the furthest point v' on γ from the initial point is within distance $2\delta + 1$ of the end of γ (i.e. the point u_2 cannot be constructed). Moreover, if one measures distance from any point $v \in Cay^1(\Gamma, \mathcal{A})$, rather than from the initial point of γ, one shows that the furthest point v' on γ from the point v lies within $2\delta + 1$ from one of the endpoints.

Gromov pointed out that a group satisfying a subquadratic isoperimetric inequality is in fact hyperbolic. Detailed proofs have been given by Papasoglu, Ol'shanskii and by Bowditch [4] (see also [6, p.422] for another presentation of Bowditch's proof). Another proof, using asymptotic cones, is given in Theorem 5.2.2 of Tim Riley's notes.

Chapter 5

Free Nilpotent Groups

The aim here is to give a lower bound for the isoperimetric inequality for free nilpotent groups, following [3]. The basic idea is to use a different method of estimating the area of a word in $\langle\langle R\rangle\rangle$. The method used has connections with group homology, but none of that theory is necessary in the constructions. Modulo a couple of elementary properties of nilpotent groups, complete proofs are given here.

When $\mathcal{P} = \langle \mathcal{A} \mid R \rangle$ is a finite presentation of the group Γ, and $Cay^1(\Gamma, \mathcal{A})$ is the Cayley graph, $\mathcal{R} = \langle\langle R\rangle\rangle$ can be identified with the fundamental group of $Cay^1(\Gamma, \mathcal{A})$. When Γ is not a finite group, this is an infinitely generated free group. We are interested in words $w \in F(\mathcal{A})$ such that $w \in \langle\langle R\rangle\rangle$, and thus there are conjugating elements p_i and relators $r_i \in R^{\pm 1}$ such that $w = \prod_{i=1}^{M} p_i r_i p_i^{-1}$. Recall that the area of w is the minimum such M.

Perhaps $\mathcal{R}$ is too complicated to be usable in computations. If we were simply to abelianise $\mathcal{R}$ and consider $\mathcal{R}/[\mathcal{R}, \mathcal{R}]$, then we would still be dealing with an infinitely generated group. If, however we consider $\mathcal{R}/[\mathcal{R}, F]$, then we are considering a finitely generated abelian group. In this quotient, $r = prp^{-1}$ for all $r \in R$ and all $p \in F$, so the number of relators in $\mathcal{R}$ is an upper bound for the number of generators of $\mathcal{R}/[\mathcal{R}, F]$.

In the world of group homology, the exact sequence $1 \to \mathcal{R} \to F(\mathcal{A}) \to \Gamma \to 1$ leads to an exact sequence

$$
\begin{array}{ccccccccc}
0 \to & \mathcal{R} \cap [F, F]/[\mathcal{R}, F] & \to & \mathcal{R}/[\mathcal{R}, F] & \to & F/[F, F] & \to & F/\mathcal{R}[F, F] & \to 0 \\
& \| & & \| & & \| & & \| & \\
0 \to & H_2\Gamma & \to & \mathcal{R}/[\mathcal{R}, F] & \to & H_1 F & \to & H_1\Gamma & \to 0
\end{array}
$$

and noting that $H_1 F$ is a free abelian group, we see that $\mathcal{R}/[\mathcal{R}, F] \cong H_2\Gamma \oplus \mathbb{Z}^k$ for some k. (In fact it is only the $H_2\Gamma$ part which is of interest to us, as is explained in [3]).

Now define a centralized isoperimetric inequality by measuring minimality in $\mathcal{R}/[\mathcal{R}, F]$. Define the centralized area of w to be $area_{\mathcal{P}}^{cent}(w) = \min\{M \mid w \in \prod_{i=1}^{M} p_i r^{e_i} p_i^{-1}[\mathcal{R}, F]\}$. Thus we count just the number of times, with sign, that each relator occurs in a product of conjugates, ignoring the conjugating element involved.

Lemma 5.1. *Let* $\mathcal{B} = \{y_1, \ldots, y_m\}$ *be any finite set of generators for the abelian group* $\mathcal{R}/[\mathcal{R}, F]$, *and let* $\ell_\mathcal{B}(w)$ *be the minimal length of a word in these generators representing the element* $w[\mathcal{R}, F]$ *of* $\mathcal{R}/[\mathcal{R}, F]$, *and let* $K = \max\{\ell_\mathcal{B}(r) \mid r \in R\}$.

(1) *Then* $\ell_\mathcal{B}(w) \leq K.area_\mathcal{P}^{cent}(w) \leq K.area_\mathcal{P}(w)$.

(2) *There is a positive constant* C *such that if* $w[\mathcal{R}, F] = y^m[\mathcal{R}, F]$ *and* $y[\mathcal{R}, F]$ *(and* $w[\mathcal{R}, F]$*) has infinite order in* $\mathcal{R}/[\mathcal{R}, F]$, *then* $m \leq C.area_\mathcal{P}^{cent}(w)$.

Proof. (1) Write $w = \prod_{i=1}^{M} p_i r_i^{\epsilon_i} p_i^{-1}$ for some appropriate choices $p_i \in F$, $\epsilon_i = \pm 1$ and $r_i \in R$, with $M = area_\mathcal{P}(w)$. Removing the conjugating elements, we have $w[\mathcal{R}, F] = \prod_{i=1}^{M} r_i^{\epsilon_i}[\mathcal{R}, F]$. Also, if the centralized area is $area_\mathcal{P}^{cent}(w) = m$, then there are $q_j \in F$, $\beta_j = \pm 1$, and $s_j \in R$ such that $w[\mathcal{R}, F] = \prod_{j=1}^{m} q_j s_j^{\beta_j} q_j^{-1}[\mathcal{R}, F]$, and removing the conjugating elements we get $w[\mathcal{R}, F] = \prod_{j=1}^{m} s_j^{\beta_j}[\mathcal{R}, F]$. Then $\ell_\mathcal{B}(w) \leq \sum_j^m \ell_\mathcal{B}(s_j) \leq Km \leq KM$.

(2) As $\mathcal{R}/[\mathcal{R}, F]$ is a finitely generated abelian group, it is a direct sum of its torsion subgroup T and a free abelian group $\mathbb{Z}^k$. Choose a generating set $\mathcal{B}$ for $\mathcal{R}/[\mathcal{R}, F]$ consisting of a generating set for T and a basis for the $\mathbb{Z}^k$ summand. Mapping $\mathcal{R}/[\mathcal{R}, F]$ onto the $\mathbb{Z}^k$ summand, $w[\mathcal{R}, F] = y^m[\mathcal{R}, F]$ maps onto an m-th power, which is non-zero if $y[\mathcal{R}, F]$ has infinite order. Thus $m \leq \ell_\mathcal{B}(y^m) = \ell_\mathcal{B}(w) \leq K.area_\mathcal{P}^{cent}(w)$. □

The point now is that in certain groups it is possible to find words in F which are very short, but represent elements of $\mathcal{R}$ which are large powers in $\mathcal{R}/[\mathcal{R}, F]$. This is easy to do in nilpotent groups, as follows.

The lower central series of a group Γ is the sequence of groups $\Gamma_1 = \Gamma, \Gamma_2 = [\Gamma_1, \Gamma], \ldots, \Gamma_{k+1} = [\Gamma_k, \Gamma]$. A group is *nilpotent* if for some k, $\Gamma_k = 1$ (of class c if $\Gamma_c \neq 1$ and $\Gamma_{c+1} = 1$). Thus an abelian group is nilpotent of class 1. The simple k-fold commutators of Γ are those commutators of the form $[[\ldots [g_1, g_2], g_3], \ldots, g_k]$, which clearly lie in Γ_k. It is not hard to show by induction that if $X = \{x_1, \ldots, x_t\}$ is a set of generators for Γ, then the classes of the simple k-fold generators of the form $[[\ldots [\zeta_1, \zeta_2], \zeta_3], \ldots, \zeta_k]$ with $\zeta_j \in X$ generate Γ_k/Γ_{k+1} (see [21, 5.4]).

The free nilpotent group of class c on n generators is F/F_{c+1} where F is a free group on n generators.

Consider $\Gamma = F/\mathcal{R}$ with $\mathcal{R} = F_{c+1} = [F_c, F]$. Then $[\mathcal{R}, F] = F_{c+2}$, and $\mathcal{R}/[\mathcal{R}, F] = F_{c+1}/F_{c+2}$. According to the general result, this group is generated by the simple commutators.

We need the following basic facts about free nilpotent groups:

The c-fold simple commutator $[[\ldots [[a, b], a], a], \ldots, a]$ is a non-trivial element in F_c, and in F_c/F_{c+1} the commutator identities give

$$[[\ldots [[a^k, b^k], a^k], \ldots], a^k] = [[\ldots [[a, b], a], a], \ldots, a]^{k^c} \mod F_{c+1}.$$

For instance, the case of ordinary commutators:

$$[a^k, b] = a^k ba^{-k}b^{-1} = a^{k-1}ba^{-(k-1)}b^{-1}(ba^{k-1}b^{-1}aba^{-k}b^{-1})$$
$$= a^{k-1}ba^{-(k-1)}b^{-1}(ba^k b^{-1}a^{-1}(aba^{-1}b^{-1})aba^{-k}b^{-1})$$
$$= [a^{k-1}, b][a, b] \mod F_3.$$

By induction it follows that $[a^k, b^k] = [a, b^k]^k = [a, b]^{k^2} \mod F_3$. The general case is similar.

Thus, returning to our example of $\Gamma = F/\mathcal{R} = F/F_{c+1}$, we have the $c + 1$-fold commutator $w_k = [[\ldots [[a^k, b^k], a^k], \ldots, a^k]$ is an element of F_{c+1} and in $F_{c+1}/F_{c+2} = \mathcal{R}/[\mathcal{R}, F]$ this is a k^{c+1} power.

Thus the above lemmas on centralized area functions give $area^{cent}(w_k) \geq Ck^{c+1}$, while $\ell(w_k) \leq 2^{(c+1)}k$. But $area(w_k) \geq area^{cent}(w) \geq C'\ell(w_k)^{(c+1)}$ and so we have obtained a lower bound for the isoperimetric inequality which is polynomial of degree $c + 1$ for the free nilpotent group of class c.

Chapter 6

Hyperbolic-by-free groups

As an example of how details of the structures of diagrams can help to give an interesting result, we look at N. Brady's result that there is a hyperbolic group containing a finitely presented non-hyperbolic subgroup. This example is a cyclic extension $1 \to K \to \Gamma \to \mathbb{Z} \to 1$. In this chapter we show that in examples of this type the kernel group K satisfies a polynomial isoperimetric inequality, following [15]. That is:

Theorem 6.1. *Let Γ be a split extension of a finitely presented group K by a finitely generated free group F, so one has the short exact sequence*

$$1 \to K \to \Gamma \to F \to 1.$$

If Γ is a hyperbolic group, then K satisfies a polynomial isoperimetric inequality.

The proof generalises easily to give an analgous result for groups satisfying a quadratic isoperimetric inequality — for details see [15]. The method of proof is to carefully study the form of van Kampen diagrams, using the area and intrinsic radius (see below) of a diagram over a presentation for Γ for a relation of K, viewed a relation of Γ, to give a diagram of bounded area over a presentation of K.

We need here the concept of *radius* of a diagram D, which is the maximum, over all vertices of D, of the number of edges in a shortest path in the 1-skeleton of D to the boundary δD. Properties of this function of diagrams are developed in section 5.2 of Tim Riley's notes. The important lemma we need is:

Lemma 6.2. *Let $\mathcal{P} = \langle \mathcal{A}; R \rangle$ be a finite presentation of a hyperbolic group Γ. Then there are constants $A, B > 0$ such that, for any relation $w \in F(\mathcal{A})$ with $\ell(w) \geq 1$, there is a van Kampen diagram over $\mathcal{P}$ of area at most $A\ell(w)(\log_2(\ell(w)) + 1)$ and of radius at most $B(\log_2(\ell(w)) + 1)$.*

Proof. Consider a relation $w = c_1 \ldots c_M \in F(\mathcal{A})$ in Γ. Draw a circle in the plane, and subdivide into M vertices labelled by integers $i = 0, 1, \ldots, M - 1$, which we consider as representatives for their equivalence classes mod n. Map this circle to a loop in the Cayley graph $Cay^1(\Gamma, \mathcal{A})$ based at the identity vertex via the word w.

In the plane, join the vertices 0 and $[M/2]$ (the integer part of $M/2$) by a straight line, and extend the map to the Cayley graph over this arc by sending this arc to a geodesic γ_1 joining the appropriate vertices in $Cay^1(\Gamma, \mathcal{A})$.

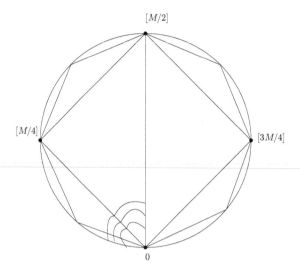

Figure 6.1: The first few subdivision triangles.

For each integer $j = 2, \ldots, [\log_2(M) + 1]$, and for each $i = 1, \ldots, 2^j$, choose geodesics in $Cay^1(\Gamma, \mathcal{A})$, the *level j geodesics*, to label the straight lines joining the vertices $[(i-1)M/2^j]$ and $[iM/2^j]$ (some of these geodesics may degenerate to points for the last j). The level j triangles are then the geodesic triangles T_j^k, for $k = 1, \ldots 2^{j-1}$, with vertices $[2(i-1)M/2^j]$, $[(2i-1)M/2^j]$ and $[2iM/2^j]$, and sides consisting of two level j geodesics $\gamma_j^{2i-1}, \gamma_j^{2i}$ and a level $j-1$ geodesic γ_{j-1}^i. At the final level take the edges in the loop w for the geodesics; at this level some of the triangles may degenerate. Notice that for each j, the sum of the lengths of the level j geodesics is at most M.

Suppose that K is δ-hyperbolic with respect to this presentation, so that each geodesic triangle can be decomposed into three triangles of area at most $\delta + 2$, a collection of rectangles of perimeter $2\delta + 2$, and a single central region of perimeter at most $3\delta + 3$. The number of these regions is at most half the perimeter of the triangle.

Filling in each of the level j triangles with these small triangles, rectangles and other central regions, construct a van Kampen diagram for the word w of area at most $A\ell(w)(\log_2(\ell(w)) + 1)$, where A is the maximum area of a minimal van Kampen diagrams over $\mathcal{P}$ for the relations of length at most $3\delta + 3$. If B' is the maximum radius of the minimal van Kampen diagrams over $\mathcal{P}$ for the relations of length at most $3\delta + 3$, then the radius of the constructed diagram is at most $\delta(\log_2(\ell(w)) + 1) + B' \leq B(\log_2(\ell(w)) + 1)$ for some B. $\square$

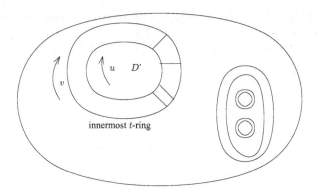

Figure 6.2: A diagram over $\mathcal{P}_\Gamma$ with some t-rings.

In the same way, it is not hard to see how (see for instance Theorem 2.3.4 of Tim Riley's notes) the fellow traveller property for synchronously (respectively asynchronously) automatic groups gives constants $A, B > 0$ (resp. $C > 1$, $D > 0$) such that each relation w has a van Kampen diagrams of area at most $(A\ell(w))^2$ (resp. $C^{\ell}(w)$) and radius at most $B\ell(w)$ (resp. $D\ell(w)$).

Proof of the theorem. For simplicity we give the details for a cyclic extension. Fix $1 \to K \to \Gamma \to \mathbb{Z} \to 1$ a split extension, defined by the automorphism ϕ of K, and let $\mathcal{P}_K = \langle \mathcal{A} \mid R \rangle$ be a finite presentation for K, where $\mathcal{A} = \{a_1, \ldots, a_n\}$. It is clear that Γ has a presentation as an HNN extension with base group K and stable letter t, $\mathcal{P}_\Gamma = \langle a_1, \ldots, a_n, t \mid R, \{t^{-1}a_i t = w_\phi(a_i)\} \rangle$. In the general case, $\mathbb{Z}$ is replaced by a free group on k generators, and Γ is an HNN extension with k stable letters and k associated homomorphisms.

Let $\Phi : \mathcal{A}^* \to \mathcal{A}^*$ be the semigroup automorphism induced by ϕ restricted to $\mathcal{A}$. As ϕ is an automorphism, there is a semigroup homomorphism (acting as an inverse at the group level) $\Psi : \mathcal{A}^* \to \mathcal{A}^*$ such that $\Psi \cdot \Phi(a_i) =_K a_i$. For each of these, choose a van Kampen diagram $D_i, i = 1, \ldots, n$ over $\mathcal{P}_K$. To complete the

proof of the main theorem, it remains to show how to obtain a diagram over $\mathcal{P}$ for a relation $w \in F(\mathcal{A})$ over $\mathcal{P}_\Gamma$ from a diagram over $\mathcal{P}_\Gamma$. As in the proof of Collins' Lemma (Lemma 2.10), the faces of the diagram over $\mathcal{P}_\Gamma$ corresponding to relations of the form $t^{-1}a_i t = w_\phi(a_i)$ combine to form annuli which we call t-rings, and fat arcs called t-corridors, meeting the boundary in t-edges.

Let $w \in F(\mathcal{A})$ be a relation over the presentation $\mathcal{P}_K$ for K. Then w is also a relation over the presentation $\mathcal{P}_\Gamma$ for Γ. Let D be a van Kampen diagram for w over $\mathcal{P}_\Gamma$. As there are no occurrences of t in w, the faces of D coming from the relations of the form $t^{-1}a_i t = \phi(a_i)$ form rings: there are no t-corridors.

Consider an innermost t-ring: i.e. inside the diagram D, there is a t-ring, i.e. an annulus A of adjacent faces all labelled by relators of the form $t^{-1}a_i t = \phi(a_i)$ such that the component D' of the complement which does not meet ∂D (the inner component) contains no relators $t^{-1}a_i t = \phi(a_i)$. Then D' is a diagram over $\mathcal{P}_K$ for the label u on the inner side of the annulus A. Let v be the label on the outer side of this annulus (the words u and v may be unreduced). There are now two cases to consider: either $v = \Phi(u)$ or $u = \Phi(v)$.

First note that applying the semigroup homomorphism Φ to the relator $r \in R$ gives a relator $\Phi(r)$. Let α be the maximum of the area of a mimimal diagram for $\Phi(r)$. In the same way there is a diagram of area at most β for each relation $\Psi(r)$.

Claim: There is a van Kampen diagram for v over $\mathcal{P}_K$ of area $\leq \max\{\alpha, \beta\} area_{\mathcal{P}_K}(u)$.

Case 1: $v = \Phi(u)$. Subdivide each edge of the diagram D' for u, such that each edge which was originally labelled a_i is now labelled $\phi(a_i)$. Each face which was labelled $r_j \in R$ is now relabelled $\Phi(r_j)$, and each of these faces can be filled in by a diagram over $\mathcal{P}$ of area at most α.

Case 2: $u = \Phi(v)$. Then in Γ, we have $v =_\Gamma \Psi(u)$. Subdivide and relabel as in case 1, but now each a_i-edge is relabelled $\Psi(a_i)$. Each face which was originally labelled r_j is now labelled $\Psi(r_j)$, and each of these can be filled in by a diagram of area at most β. This diagram for $\Psi(u)$ can be made into a diagram for v as follows. Noting that $u = \Phi(v)$, we have $\Psi(u) = \Psi \cdot \Phi(v)$ and that if $v = c_1 \ldots c_p$ with $c_j \in \mathcal{A}$, then $\Psi \cdot \Phi(v) = \Psi \cdot \Phi(c_1) \ldots \Psi \cdot \Phi(c_p)$, it suffices to add diagrams for each relation $c_j =_K \Psi \cdot \Phi(c_j)$ If γ is the maximum area of these diagrams, then there is a van Kampen diagram for v over $\mathcal{P}$ of area at most $\beta\, area_{\mathcal{P}}(u) + \gamma\ell(u) \leq \gamma'\ell(u)$.

To obtain a bound on the area of a $\mathcal{P}_K$ diagram for w it suffices to note that t-rings can be enclosed to a depth of at most the radius of D, and removing innermost t-rings one after the other multiplies area by at most $C = \max\{\alpha, \gamma'\}$.

Thus, as the original diagram D over $\mathcal{P}_\Gamma$ can be chosen of area at most $A\ell(w)(\log_2(\ell(w))+1)$, and of radius at most $B\log_2(\ell(w)+1)$, there is a $\mathcal{P}_K$ diagram for w of area at most $C^{B\log_2(\ell(w)+1)} A\ell(w)(\log_2(\ell(w))+1)$ which is bounded by a polynomial function of $\ell(w)$. $\qquad\Box$

Bibliography

[1] J. Alonso. Inégalités isoperimétriques et quasi-isométries. *C. R. Acad. Sci. Paris Série 1* **311** (1990), 761–764.

[2] J. Alonso, T. Brady, D. Cooper, V. Ferlini, M. Lustig, M. Mihalik, M. Shapiro, and H. Short. Notes on word hyperbolic groups. In *Group Theory from a Geometrical Viewpoint*, ed. by E. Ghys, A. Haefliger, and A. Verjovsky, pp. 3–63, World Scientific, 1991.

[3] G. Baumslag, C.F. Miller and H. Short. Isoperimetric inequalities and homology of groups. *Invent. Math.* **113** (1993), 531–560.

[4] B.H. Bowditch, A short proof that a subquadratic isoperimetric inequality implies a linear one. *Michigan Math. J.* **42** (1995), 103–107.

[5] N. Brady. Branched coverings of cubical complexes and subgroups of hyperbolic groups. *Jour. London Math. Soc.* **60** (1999), no. 2, 461–480.

[6] M.R. Bridson and A. Haefliger. *Metric spaces of non–positive curvature.* Springer-Verlag, 1999.

[7] A. Cayley. On the Theory of Groups. *Proc. L.M.S.* **9** (1878), 126–133.

[8] I. Chiswell, D. Collins and J. Huebschmann, Aspherical group presentations. *Math. Z.* **178** (1981), no. 1, 1–36.

[9] M. Coornaert, T. Delzant and A. Papadopoulos. *Notes sur les groupes hyperboliques de Gromov.* Lecture Notes in Mathematics 1441, Springer-Verlag, 1990.

[10] M. Dehn. Über unendliche diskontinuierliche Gruppen. *Math. Ann.* **71** (1912), 116–144. Translated by J. Stillwell: On infinite discontinuous groups. In *Papers on Group Theory and Topology.* Springer-Verlag, 1987.

[11] R. Fenn. *Techniques of Geometric Topology.* Cambridge Univ. Press, London Math. Soc. Lecture Notes No. 57, 1983.

[12] R. Fenn and C.P. Rourke. Klyachko's method and the solution of equations over torsion free groups. *L'Enseignement Mathématique* **42** (1996), 49–74.

[13] E. Ghys. *Les groupes hyperboliques.* Séminaire Bourbaki 722, S.M.F., 1990.

[14] E. Ghys and P. de la Harpe. *Sur les groupes hyperboliques d'aprés Mikhael Gromov*. Birkhäuser, Progress in Mathematics 83, 1990.

[15] S.M. Gersten and H. Short. Some isoperimetric inequalities for kernels of free extensions. Proceedings of the Congress in honour of John Stallings. *Geométrica Dedicata* **92** (2002), 63–72.

[16] L.I. Greendlinger and M.D. Greendlinger. On three of Lyndon's results about maps. In *Contributions to group theory*, ed. by K.I. Appel, P.E. Schupp and J. Ratcliffe. Contemp. Math. 33, 212–213, A.M.S, 1984.

[17] A. Hatcher. *Algebraic Topology*. Cambridge Univ. Press, 2002. Also (legally) available at
http://www.math.cornell.edu/~hatcher/AT/ATpage.html.

[18] D.R. Holt and K. Röver. On real-time word problems. *J.L.M.S.* **67** (2003), 289–301.

[19] A. Klyachko. Funny property of sphere and equations over groups. *Comm. in Alg.* **21** (1993), no. 7, 2555–2575.

[20] R. Lyndon and P.E. Schupp. *Combinatorial group theory*. Springer-Verlag, 1977.

[21] W. Magnus, A. Karrass and D. Solitar. *Combinatorial Group Theory*. 3rd edition, Dover, 1976.

[22] C.F. Miller III and P.E. Schupp. The geometry of HNN extensions. *Comm. Pure Appl. Math.* **26** (1973), 787–802.

[23] J. Milnor. A note on curvature and the fundamental group. *J. Diff. Geom.* **2** (1968), 1–7.

[24] W.D. Neumann and M. Shapiro. *Geometric group theory*. Notes for the ANU short course, 1996. Topology Atlas http://at.yorku.ca/i/a/a/i/13.htm.

[25] Y. Ollivier. Sharp phase transition theories for hyperbolicity of random groups. *Geom. Func. Anal.* **14** (2004), 595–679.

[26] Y. Ollivier. On a small cancellation theorem of Gromov. *Bull. Belgian Math. Soc.*. To appear.

[27] A.Yu. Olshanskii. Hyperbolicity of groups with subquadratic isoperimetric inequality. *Intern. J. of Algebra and Comput.* **1** (1991), no. 3, 281–289.

[28] A.S. Švarc. Volume invariants of coverings. *Dokl. Akad. Nauk. SSSR* **105** (1955), 32–34.

[29] E. van Kampen. On some lemmas in the theory of groups. *Amer. J. Math.* **55** (1933), 261–267.

[30] C.M. Weinbaum. The word and conjugacy problem for any tame prime alternating knot. *Proc. Amer. Math.Soc.* **22** (1971), 22–26.